"如果一艘船不知道该驶去哪个港口,那么任何方向吹来的风都不会是顺风。"
——摘自卢西奥·阿耐奥·塞内加(Lucius Annaeus Seneca)的《给 Lucillius 的信》第 71 封。

致 Nemi, Alessandro, Spartaco

托尼诺·帕里斯
Tonino Paris

托尼诺·帕里斯（Tonino Paris），1946 年出生，建筑师、评论家和教授。在罗马大学（Sapienza University of Rome）担任教授直至2016年，如今在博洛尼亚大学（University of Bologna Alma Master）任职。1973 至 1996 年间，创办并领导 BMP srl 建筑师事务所，在不同的设计领域开展业务。

曾在多个学术机构任职，其中包括意大利文化遗产部国家设计委员会主席，国立大学委员会成员，罗马大学 ITACA 系和城市规划、设计与建筑技术系主任。1994 年，创办并领导罗马大学设计专业，同时创办并领导设计工厂实验室（Sapienza Design Factory Lab），在此开展了旨在建筑与设计领域的工艺和产品创新相关的研究及设计活动。

2012 年，创办并管理推广设计文化的学科期刊——DIID,Disegno Industriale _Industrial Design。

作为作者曾出版发行多本书籍专著：

2015	Le Corbusier_antologia di scritti
2013	A Design selection
2011	Il design italiano 20.00.11
2010	Where design grows up
2009	09 young design. European Creativity_DAS 09
2009	Designer's: Exhibit, Product, graphic, Fashion and food
2005	Made in italy | Il design degli italiani | Made in Italy_ Italians' design
2004	Design alla Sapienza: 1994-2004
2004	YounGeneration, prodotti e prototipi
2000	Prodotti e progetti: 1995-2000

托尼诺·帕里斯 著

林晶晶 译

设　计
文 本 与 语 境

华东师范大学出版社

设计 _ 文本与语境

作者
托尼诺·帕里斯
林晶晶　译

在此特别感谢魏劭农教授，没有他的真诚参与，这本书将无法发行。他对作品内容、形式的不断完善，对于这本书的出版起到决定性的作用。其次，还要特别感谢林晶晶，感谢她对翻译工作付出的耐心及牺牲。

同时，我还要感谢那些多年来共同参与Disegno Industriale_Industrial Design（以下简称《DIID》）杂志工作和致力于传播设计文化的同事们。正是他们积极地参与，让我产生了更多的想法和思考，逐渐形成了设计学科的理解方式。我认为，设计学科是一件工具，借助它我们可以进行学习并提出想法，来帮助改善我们的生活以及社会关系。

在他们当中，我要特别感谢一众与我有着数年友谊的教师：文森佐·克里斯达洛（Vicenzo Cristallo），塞西莉亚·塞克奇尼（Cecilia Cecchini），范德利克·达·伐柯（Federica Dal Falco），洛雷达娜·蒂鲁克（Loredana Di Lucchio），罗勒佐·尹百斯（Lorenzo Imbesi），萨布里娜·鲁奇百洛（Sabrina Lucibello），卡尔洛·玛乐狄诺（Carlo Martino）。最后，我要感谢罗马大学设计学院的所有师生，在这所由我创办并为之奉献一生的设计学院里，我收获良多。

序　言

早在上个世纪的二三十年代，随着城市现代化商业的发展，在中国的一些学校中已经出现了与设计相关的一些教学科目，例如广告、海报和工艺美术等等，但是作为独立学科的设计学系统的教学却是在上世纪的八十年代中期才开始逐渐的出现在中国的大学之中，这与我开始在大学中从事艺术教育工作的时间恰好吻合，而我自己真正有机会去深入的观察和了解这门学科还是在十多年前我开始在大学中担任设计学院院长之后的事情了。

今天在中国的大学中每年都有超过二十万以上的学生在学习设计，这是一个相当惊人的数字，而我们面临的问题是，由谁来教这些学生。学习设计并非易事，而设计教学就更是一项非常困难和具有挑战性的工作了。很显然我们根本无法在短期内找到那么多合格的教师来从事设计的教学工作，我们甚至很难找到合适的教材供学生阅读，而目前正在使用的许多教材不仅不能正确引导学生真正地理解设计，相反还会对学生产生各种严重的误导，这是一件非常令人担忧的事情，文本和语境在我们学习和讨论与设计相关的问题时至关重要。

六年前我们华东师范大学设计学院在意大利罗马举办了学生作品展，展览期间我在罗马一大的校园里第一次见到了时任罗马一大建筑系主任的托尼诺·帕里斯教授，从那以后我们和罗马一大建筑系及帕里斯教授本人一直保持着非常密切的学术与教学上的合作与交流。多年前帕里斯教授创办并主编了《设计工业·工业设计》（DIID）杂志，用他自己的话来说，就是希望能够同时站在历史和当下的角度来重新定位"设计的真实尺度"。DIID杂志自2002年创刊以来，共有超过450位意大利及全球设计专业领域的专家和学者发表了850多篇专业论文，每一期杂志论述一个特定的主题，并对这一主题进行了发展性的描述，其内容涉及工业生产过程的发展和创新、设计实践的发展历程、应用艺术及设计语言的研究、设计实践及理论的实验性演变、物质与非物质的人工造物发展历史研究，以及技术、科学、社会、经济在人工环境建设中的协调发展，等等。对这些在设计领域中具有代表性的主题的讨论，如同围绕着产品在系统设计的过程中技术和文化的延伸所进行的一场世俗性的辩论，通过复合性的结果来击中移动性的目标。帕里斯教授将DIID作为对设计领域研究的实用性工具，试图在人类活动的时间和空间范围中去发现和讲述关于设计的故事以及设计本身的定义，将设计作为"人类自身活动的一种延伸"，通过帮助读者对隐含在设计中的物质和寓意的理解，来定义这种"延伸"，从而诠释了设计如何将人们的需求转化为产品，来改善生活以及人与人之间的关系，并从中发现设计所具有的时代特有的意义。

两年前帕里斯教授将他刚出版的著作《文本与语境》赠送给我，我发现这是一部涵盖了帕里斯教授学术生涯特定领域的著作，总结了他对与设计相关的不同主题的思考和想法：人工造物、创新价值、社会维度和其他，同时也概括了"设计"这一词本身的含义，尤其是试图去探索设计这个词在其表层含义下的真正意义，并且对其加以维护，从其表面的含义中引申出更多深层次的内容。从某种意义上说，帕里斯教授的这部著作更是对其多年来主编的DIID杂志的一次重新的阅读、认知和评论，正是基于这个方向，帕里斯教授确定了这部著作的主要内容：文本与语境，以文本摘录的形式来加以呈现，他将杂志的一系列选段依照其个人的观察点和兴趣重新进行了分类和评述，这或许与他仍在从事的教学工作有关，它如同一个超级的网络文本，为了使这些看上去似乎有些碎片化的文字和段落更加具有可读性，帕里斯教授将它们重新编入到一个由许多"设计话题"构成的特定框架中。这些话题构成了本书第一部分"设计人工造物及相关主题"，在这部分内容中，帕里斯教授以"第一人称"的身份来进行叙述。而后面的"摘录"部分的内容则主要来自与他共事多年的教授们的文章和讲座。在结尾部分帕里斯教授对纯粹案头式的设计文化进行了批判性的论述，同时提出了与当代人工造物相关的设计实践的方法论，并且阐述了如何通过实践操作将理论、方法与材料本身进行对比和研究。

无论是对于帕里斯教授个人的学术生涯还是对于当代的设计与设计教育界来说，本书都是一部重要的文献著作。帕里斯教授试图用其设计的关于设计的话题来唤起关于设计的讨论和设计工作本身所引发的情感、设计对于未来人与环境之间的变化关系所产生的一系列思考，以及关于设计的永恒性的主题。由这部著作所引发的一系列关于设计的批判性的思辨，不仅表达了关于设计的可持续性的辩证对话，同时也表达了帕里斯和他的同事们对于设计文化的热爱，并不断地激发起将设计文化和对这种文化的热爱传递给年轻一代的热情。

显而易见，这部著作的读者群体包括了所有对设计和设计文化感兴趣的人，尤其是在大学中学习或教授设计的人，对他们而言，设计应该是一个实验场所，同时也是一个具有无限发展因素和可能性的世界。语言表述的逻辑和修辞是帕里斯教授著作的重要组成部分，而这一几乎与著作内容本身同等重要的部分，在翻译的过程中通常是会有所流失的，因此，我们将原著的英文版一并收入并出版，以便于读者对照阅读。感谢帕里斯教授慷慨地贡献了本书的中文版权，感谢华东师范大学出版社总编辑阮光页先生和他领导的杰出的编辑团队，感谢帕里斯教授的博士生，华东师范大学设计学院产品设计专业的青年教师林晶晶女士，她能够在不到半年的时间里完整地翻译出了这部意大利原版的学术著作实属不易。

某种程度上我们可以这样来理解这部著作在中国的出版，这是一个热爱设计的人送给另一些同样热爱设计的人的礼物。

华东师范大学设计学院院长　**魏劭农 教授**

2018 年 10 月于柏林

目　录

目 录

p.11 >　　[1]　**设计人工造物及相关主题**

p.25 >　　[2]　**文本与语境**

p.28 >　　　　**工业设计与设计**
　　　　　　无处不在
　　　　　　设计自主
　　　　　　不可避免的工业化
　　　　　　教授设计
　　　　　　物质与材料
　　　　　　教育与培养
　　　　　　相同与差异
　　　　　　展示活动
　　　　　　复杂的虚拟结构
　　　　　　它是设计
　　　　　　物品的语言
　　　　　　未来场景

p.37 >　　　　**构思消费品**
　　　　　　新型科技
　　　　　　色彩
　　　　　　教学方式
　　　　　　可持续技术
　　　　　　连接
　　　　　　含糊的概念
　　　　　　教育经历
　　　　　　市场需求
　　　　　　惊艳亮相
　　　　　　沟通交流
　　　　　　错觉感
　　　　　　有关材料的思考
　　　　　　磁卡
　　　　　　有益的陈旧
　　　　　　俏皮亮相

p.46 > 事物、物品与产品
　　　　　复杂流程
　　　　　从未间断
　　　　　激发创造力
　　　　　建筑物品
　　　　　战略价值
　　　　　为运动设计
　　　　　无名但工业化
　　　　　物品与环境之间
　　　　　真实性与模糊性
　　　　　技术创新
　　　　　杂交产品
　　　　　设计博物馆
　　　　　升级

p.53 > 伦理与美学
　　　　　工具箱
　　　　　疑问
　　　　　设计师的培养
　　　　　智能文化之争
　　　　　homo faber
　　　　　演出
　　　　　陷入混乱
　　　　　1个H，4个W
　　　　　细微或显著
　　　　　理由
　　　　　高科技美学
　　　　　系统共存
　　　　　伦理&思想
　　　　　主导品位

p.63 > 有用与无用
　　　　　终身教育
　　　　　记忆

建筑的危机
可整合设计
新的平衡
大型农场
与人互联
实践
食品设计
认知行程
稀有且独特
一只章鱼

p.72　>　　　混合现实
学术课程
相互影响
交互设计
混合过程
电影设计
社会本质
转移
微型设计

p.79　>　　　文化创新
色彩社交
语义价值
设计院校
结合
生产领域
意大利风格
意大利印象

p.87　>　　　传统与变革
一堂课
建筑元件
新的刺激

　　　　　　　　　　　　　地中海 —— 两地之间
　　　　　　　　　　　　　大众设计
　　　　　　　　　　　　　美化
　　　　　　　　　　　　　无名
　　　　　　　　　　　　　产品流
　　　　　　　　　　　　　复杂的文化

p.97　　>　　　　[3] **摘录**

p.99　　>　　　　设计命名
　　　　　　　　　　　　　诠释
　　　　　　　　　　　　　人工造物
　　　　　　　　　　　　　物品传记
　　　　　　　　　　　　　创造力 —— 高频词汇
　　　　　　　　　　　　　设计的伦理
　　　　　　　　　　　　　现代与后现代设计危机
　　　　　　　　　　　　　设计文化中的材料

p.107　>　　　　设计中的标准
　　　　　　　　　　　　　称之为标准
　　　　　　　　　　　　　产品特征
　　　　　　　　　　　　　标准与现代运动

p.109　>　　　　新的范式
　　　　　　　　　　　　　开放源代码时代的设计
　　　　　　　　　　　　　桌面制造设计
　　　　　　　　　　　　　按需设计

§
[1]

设计人工造物及相关主题

§
[1]
p.13

设计人工造物及相关主题

本书以《设计人工造物及相关主题》为开篇,在此段落中,包含了对所有主题的批判性论述。这些主题形成了两大主线,它们最终相互完善。其中,一条主线与设计文化涉及的所有主题相关,另一条主线则与设计人工造物的理论方式与实践操作相关。借助完整与非完整参半的主题描述,能够唤起深刻情感的设计操作,对不变的设计物质及多变的背景进行的反思,面向近未来设计的眼光,进而以一种主观的方式进行论述。

不难看出这本书面向的读者群体是很明确的,包括对设计感兴趣的人,对于他们而言,设计是实验场所和发展因素,当然还包括那些在设计学院学习或者教授设计的人。

一直以来,在我的所有担心中,我最害怕违背职业道德,尤其是在讨论设计主题的时候。这些年来,我与学校的之间的关系变得越来越艰难,而且常常让人心生疲惫。看着教师和学生们,让我的心与学校间的微妙关系能够续存下来。他们在教室里分享有用的知识,哪怕这间"教室"是不存在的。然而何为"有用"?它是一个要求很高的词汇,特别是教授的课程很复杂且需要验证的时候,我们需要依据理论模式及实践活动教学,让那些想要设计人工世界的学生,从人工造物出发。当然,身负教师这一职责,我依然感受到,我还需要做些什么。我从一系列的评论行文中找到灵感,因而开始创作这本书籍。我认为这本书的目录相对主观,它的出发点不涉及任何指定主题,而是一种个人的"自我发挥"。这也许是唯一的方法,来平衡我广泛兴趣的叙述需求和坚定的信仰理念。虽然我的选择性论述会让读者看到碎片化的内容,但是那些对设计感兴趣的人,他们定会将设计视为实验场所和发展因素,这才是我的理想对话者,我相信,他们可以很好地理解本书内容。

• • •

环境 > 新作品的构思往往非常复杂,知识储备是开启构思首要且必要的条件。无论作品大小,首先要对设计操作所处的背景环境加以了解。然而,对于设计师来说,这个背景环境不能简单直接地与我们习以为常的自然环境相混淆。自然环境一直存在,而背景环境却是一种虚拟所在,处处充满陷阱又十分重要,我们称之为"设计的文化土壤"。提到背景环境,我不得不考虑它的"圆滑善变"。它表面上看似毫无波澜,实际上处处充斥着"欺骗"。正因为它的本质难以捉摸,所以显得非常模糊暧昧。为了更好地理解背景环境,我以后现代主义为例进行阐述,它存在的时间虽然极为短暂但却蔓延渗入文化环境当中,进而成为了我们当代社会的一个重要特征。

安伯托·艾柯(Umberto Eco)非常有效地概括了它的作用与特征:"后现代主义标志着置于世界秩序之上的"大叙述"*的危机;同时,后现代主义通过不同的方式与虚无主义相结合,致力于重新审视过

---

* 大叙述:放眼未来的哲学思想

§

[1] 设计 _ 文本与语境

去。后现代主义表现了一种从现代化到当前无名浪潮的过渡,当今特色的环境,保障了个人以灵活的方式来解决各种不同问题的可能性。一般来说,每一个集体价值诉求,会让个人感受到自己是集体的一部分,进而阐述各自的需求。当集体诉求遭遇危机,顺势而来的是意识形态危机的出现。伴随着集体概念的危机,一种不受约束的个人主义出现了,人与人之间不再是相互陪伴的路人,而成为了相互竞争的对手。这种主观主义破坏了现代化的基础,使它变得脆弱不堪,一切都以一种流动的形式溶解了。对于一个没有参照的人来说,唯一的解决方案就是不论任何代价以价值形式呈现。在这种毫无目的的贪婪中,人们为了加入这场"狂欢"而不断地放弃旧事物。那么,什么能取代这种现代化的消融呢?我们还不知道,但是这个过渡时期会持续很长一段时间。因为问题在于,政治家与大多知识分子还没有理解到这一现象带来的严重后果。"[1]

这是一篇有关现代化主题的段落节选,它应该给我们每个人都产生新的强烈的推进力量。鼓励我们重新关注我们设计对象的内容,而不是那些短暂存在的外表。因此,在做设计的时候,应该将人放在我们设计目标的中心。设计产品应该回归简单,其中必不可少的环节就是理解人们的需求,将其转化为产品,来改善人们的生活和他们的社会关系。

最近的两个引人入胜的展览,让我对于设计功能产生了一些想法。也正是在这些特殊的空间环境中,我隐约地感受到我的想法得到了近乎哲学的支持。这两个展览分别是:2014 年,第 14 届威尼斯国际双年建筑展,策展人为雷姆·库哈斯(Rem Koolhaas);2016 年,在第 21 届米兰三年展中的一个展览,名叫《新史前时代——100 个动词》,策展人为安德列亚·布兰茨(Andrea Branzi)和原研哉(Kenya Hara)。

米兰《100 个动词》展览中,表现了人类的 100 种行为。在不断的进化当中,不变的是人类总是能制作出与之对应的精确器具。因此,我们在标记着人类文明发展的创造与物品之间行进。展览中的动词用来描述一些实践操作,比如测量、航行、书写、烹饪、演奏以及杀害;反之,则是通过那些构成人类历史的物件,表达一些无形的不变的抽象感受,比如爱恋、吸引、想象、发明和思考。如果说人类的生活与愿望息息相关,进化则可以理解为对愿望的实现。有吃饭的愿望,所以学习烹饪;有交流的欲望,所以进行表达(最早由符号进行表达,或者彼此的眼神交流,慢慢地通过语言,变成借助文字。如今,使用各种媒介,交流起来更加没有障碍)。同时,人类与死亡的接触是不可避免的,因此,清单上不能缺少毁灭、刺伤、威胁等动词。这些动词唤醒了人类思想中产生的所有邪念,它们无论过去还是现在都是没有区别的。展览当中,每个动词对应的活动都阐述了涉及的性能、形式和技术的不断发展,这些变革让创造新的物品变得可能。正如安德列亚·布兰茨(Andrea Branzi)所言,在 100 个器具中滚动的 100 个动词就好像"来自黑暗的历史、无限的空间的极具生命力的神秘能量,借助科学研究,生产出能够替换人体所需的'零件'"[2],进而丰富了人类的生存方式。"对原研哉(Kenya Hara)而言,这些"零件"为人类的欲望提供了物质形式,因为人类通过这些器具发展了他们自身的能力。同时,由于这些器具人类产生

设计人工造物及相关主题

了更多的欲望，因此，人类"欲望的历史"可以称为"器具的历史"[3]。如今，我们正处于21世纪，一个全速向人工智能行进的时代，思考人类未来的发展以及未来所处的生存方式是非常重要的。反思过去，从而得知未来人类真正的需求。

由安德列亚·布兰茨（Andrea Branzi）和原研哉（Kenya Hara）策划的展览是一堂深入探索的课程，展示了homo faber* 改善物品的全部设计能力及智慧。我们不难发现，即使只是一个很小的物件，也让我们思考对人类行为进行诠释的必要性。虽然事实上，区分史前物件和当今物品的技术差异是显而易见的，但是就其设计状态而言是相同的。根据各自需求塑造物质的意愿，形成了整个人类历史的主线。即使人类的需求总是在不断地更新，材料及科学技术在不断地发展，通过发明器具来满足自身需求的能力始终不变。与此同时，这也需要那些设计师通过满足真实的需求，来最终实现务实的解决方案，而不是像那些当今的建筑师明星一样，一味地以自我为参考。

通过双年展的主题：基本元素，策展人雷姆·库哈斯（Rem Koolhaas）明确指出建筑师与建筑师明星的差别。他从基本元素出发重新回归建筑，并以此为关键词进行布展，赋予展厅以一种特殊的建筑历史博物馆的感觉。雷姆·库哈斯（Rem Koolhaas）在主场馆中展示了一系列建筑构成元件的家族化谱系及其变化。他挖掘过去的基础元素，并进一步研究当下，从而对未来有个清晰的展望。在对建筑元件的重新解读中，他似乎含蓄地传递了一种让人振奋的信息：现代设计文化在全球化压力下并没有完全消失。雷姆·库哈斯（Rem Koolhaas）要求我们克服城市形态的阻碍，也正如安伯托·艾柯（Umberto Eco）所言（从微弱思想到解构主义，从相对主义到后现代主义）这是长期以来受意识形态危机影响而生成的。即需要推动以服务人类为最终目标的建筑和设计的发展，来重新赢得大众对这一主题的兴趣。正因如此，雷姆·库哈斯（Rem Koolhaas）认为，重新学习这些用于构成房屋、建筑、城市的元素是十分必要的。地板、墙面、天花板、窗户、立面、阳台、走廊、壁炉以及楼梯，这些不可或缺的建筑DNA元素被所有建筑师广泛使用。这一回顾通过已有的建筑标准、原型和类别，以向人类提供新的模式为目标，必要时使用新型技术进行实验研究。

· · ·

标准与原型 > 物质与非物质元素，归根结底是标准与原型的概念。所谓标准，是指由结构元件形成的古老结构，尽管其形式、功能、技术在不断地演变进化，它是始终保持不变的。所谓原型，是指为了适应新的功能满足人类新的需求，而形成的新的结构。因此，标准是新的原型衍生的参考。无论标准还是原型，都可以被视为是基础元素，是建筑与产品的构成物质，是forma urbis**（一个持续五千

---

* homo faber：工具制造者。上帝创造人类之后，人类根据其需求，能很好地运用自己的智慧，来制造工具并借助工具适应并改变所处的环境。

** forma urbis：城市形态如同建筑与开放空间形成的统一体

§ [1] 设计_文本与语境

多年的混合体）的表现形式。随着周期、经济因素的变化，以及其它不同的原因，彼此间相互独立的元素将建筑和产品转化为一个复杂、古老或现代，独特或标准的混合拼贴。如果在固定结构和组织系统中对标准与原型进行再研究，可以让我们重新认识它们的文化喜好，它们被遗忘的符号，它们的技术演变，它们对气候的适应，它们的政治因素，它们的规范标准，等等。

在设计领域中，标准作为构成物质这一概念的转化并不是立即出现的。安德列亚·布兰茨（Andrea Branzi）和原研哉（Kenya Hara）将日常用品融入为动词，进而实现这种转变。人类的动作，他们参与的活动，以及他们的愿望，形成了历史中的不变因素。物品满足人类的需求，物品如同人类的身体或思想的延伸物，这并不是偶然。例如，作为防御和进攻的物品，一开始只是简单的石头，之后变得越来越复杂。

从外观或分类角度来看，一连串物品的更新换代总会让那些之前的物品显得老旧过时。事实上，无论是在它们的设计生产阶段，还是在它们的识别使用阶段，我们对人工造物的认识中有着"类型学认知"。如果缺乏这种对物品框架的基本认识，我们就好像置身于一个无法辨认的物质世界，感到手足无措。因此，我们需要让每个物品尽可能地拥有且保持自身的独特属性，即"借助特定的形式、结构和功能来与其它技术产品进行区分。设计本身决定了产品的基本特征，并通过特定的技术元件丰富其独特性。而后在任何类型产品中，尤其是在强大技术革新时代，配置结构使得它们更加具有识别性。"[4] 然而近年来，产品及其生产领域的研究表明，这种分类方式依然存在着明显的不完整性。

• • •

工艺技术与人工造物 > 从身体构造及感觉器官看来，人类是大自然中最柔弱的动物，但是却被迫着赤手空拳与大型食肉动物较量，这也注定了人类要在日常生活中挣扎求生。以视觉为例，我们发现人类和大多动物的视觉灵敏度完全没有可比性，更别说在黑暗中的适应能力了。

有些猛兽的可视距离是人类的十倍甚至百倍，更别说它们对行进路径和速度的判断力，这一技能在它们的狩猎过程中尤为突出。当猛兽看见猎物，它们开始评估接近猎物的时间，考虑追捕的距离，来判断捕获猎物的可能性。尽管如此，赤手空拳的人类运用自身的智慧依然能够战胜大多数的动物。人类设计并制作各类延伸性物品，必要时可以弥补自身的不足。棍棒是人类设想的最原始最基本的防御工具，慢慢地，越来越有用的工具逐渐被制造出来：斧头、刀、箭等，直到最终手枪的实现。同样，人类视线无法覆盖的距离，可以用眼镜、望远镜来弥补。又或者是使用肉眼无法察觉的物品，比如隐形眼镜，它甚至还可以改变虹膜的颜色。因此，从最原始需求慢慢衍生出的看似多余的渴望，同样可以使用延伸性物品来满足。

关于视觉和延伸性物品的初步讨论，简单明了地引入了工艺技术与人工造物的概念。工艺技术，在这种情况下，就是使用合适的工具与适当的方法来完成一个产品。正如，我们需要一个产品来改善我们所处

设计人工造物及相关主题

的生存环境，这种将需求转化为产品的过程是人类智慧与技能的结晶。人类借助智慧设计出满足明显需求的高性能产品；同时，借助技能选择合适的工具与操作方法来生产产品。如果在上述过程中加入了实践操作，那么产品的设计与生产就不再分离。所谓的手工产品完全符合这一典型观点。在这过程当中，"思想活动"需要储备大量的知识与相关的文化，而"制作活动"则意味着使用工具与方法，合理地将材料转化为产品，并通过一系列训练来提升这些能力。

这些产品没有自主的艺术效度，至多表达了一种集体审美意向。产品的需求并非一成不变，所以这个过程是没有极限的。也正因如此，产品总是会被进一步地改善，被付诸几代人的不断努力。以持续的传统为基础，借助不断累积的经验，技术得以发展。以物质文化产品和应用艺术产品举例，能更好地简化这个概念。比如，家用厨房器皿一直保持着它收纳、保存、烹饪食物的功能，但是随着几代人不断地更替，这些物品慢慢地被改善。

关于这个主题，我想谈谈儿童玩具，它们在一定程度上代表了技术智慧和集体审美。在意大利一个贫困区域，我看到了一个类似溜溜球的玩具，它是由一根灌木杆、一根细绳以及一片果皮制成。它没有别的名字，被称之为"Briquedo"，这是一个非常笼统的定义，仅仅是用来表示这是一个玩具而已。实际上，这是一个看似简单却很复杂的物品。玩这个玩具的时候，从各个部分受力来看，它蕴含着静力学知识、功转化为动能的概念、加速度和惯性的基本原理，以及对其构成材料极限的了解。同时，用手紧握玩具时还涉及到人体工程学的比例原则。所有的这些，不过是一代又一代人累积的经验在这个物品上得以沉淀。也许孩子们甚至不能够理解，但是可以肯定的说，这个物品蕴含了大量的技术知识，而这些知识已然成为集体的财富。

同样的经历也出现在小型自行车和其它物品上。实物被越来越精致的模拟模型所替代，它们与实物几乎没有太大差别，配件越来越丰富（也有可能改变各个部分之间的比例），颜色也越来越缤纷。作为个人玩具而诞生的物品最终成为一件商业产品，这是一种充满矛盾的高度进化的手工艺带来的结果。因为它可以在不需要设备的情况下，仅仅依靠双手，依靠个人技能进行再生产。

最后介绍一个最简单的玩具：卡车。它满足了孩子们的模仿行为，成全了他们目前不可能实现的梦想，即成为卡车的车主和驾驶者。但谁又知道，也许玩具带来最大的乐趣是生产它们而不是像这样使用它们呢？而此时的玩具最值得欣赏的就是选择了木材作为它的材料。"就结实而柔软、接触时自然而然的温热之感而言，木头都是理想的材质；它可以从各个面上削去过尖的角度；孩子搬弄它，碰撞它，它不会震动，吱嘎作响，出声却是低沉而清脆。这是种亲切而蕴满诗意的材质，让孩子仿佛与树木与自然接触一般。木头不伤人，也不易被弄坏；它不会破碎，只是被磨损、用旧，可持续用很久，伴随着孩子的成长，渐渐改变了外在物品与手的关系；即使它寿命到了，也只是逐渐缩小，并不是呈现膨胀的状态，不像那些机械玩具因弹簧坏了而局部鼓胀起来。木头制成了蕴含物本性的物品，永恒的物品。"[5]

§

[1] 设计 _ 文本与语境

当概念方案和执行方案相结合时,可以重新生产在技术领域和构成实践方面具有代表性的产品,借此来推广所有人都能接受的普遍经验。如果概念方案和执行方案分离,那么,就存在设计师及其设计方案。在产品的实现过程中,实际上逐渐依靠生产或机械,因此不存在工人的自主操作:设计方案以产品的诞生为结束。在设计方案中,设计方案定义所有,首要的就是产品的形式与功能,其次是生产执行过程,但并不会引入其它审美价值。其中还包含了实际执行,也就是评估在规定时间内产品技术得以实现的可能性。

后工业文明到工业文明的过渡,使得产品的生产及其技术得以评估。更为明显的是在后工业时代的文明与交流中,人们认为起决定作用的是工具而不是思想。在设计初始阶段,设计师的作用是明确的,他们通过研究成果与科学发现选择合适的技术与执行步骤,以完善产品的性能来尽可能地满足人类的所有需求。这一需求不会以功能性得以满足而终止。

与此同时,产品的交流价值、美学内容、象征符号以及展示需求,即使它们是多余的,也依旧受到越来越多的关注。然而这些价值大大缩短了产品使用时效,更进一步加速了产品的淘汰。

为了让人类在恶劣危险的大自然中存活下来,延伸性产品应运而生。它们就像是功能性设备,具有高科技含量及高度沟通性。这些人工造物有时甚至还需要响应人类看似多余的需求。

以技术产品为例,它们的诞生来源于设计师的决策。我们可以想一想用来指明时间的物品:时钟。测量时间是人类在早期需要掌握的一个非常重要的技能。人类以此来知晓白天黑夜的距离、月相盈亏的周期、四季的更迭、生命的孕育,生存于大地之上,主宰经济社会。为此,人类利用阴影设计完成了测量时间的器具;又或者通过重新解读材料位置的改变来获取时间信息,比如沙漏。慢慢地,逐渐完善钟表的技术含量,将机械原理应用于发条钟表或者挂钟,直到实现绝对精度。很长一段时间里,人们选择钟表,看中的是其准确性或设计者的知名度,外观反而不是那么重要。

直到电子技术的出现,随着印刷电路(在操作电子设备的过程中利用工业加工印压的微小且成本低廉的部件)的普及,这一精确的学科让钟表的技术含量变得无人能及,不仅在精度上无法超越,甚至在制造成本上还降低不少。设计替代一切,重新定义了物品的价值:外观品质及其象征意义。

斯沃琪(Swatch)手表就是典型的标志。正是它的"原创设计",让它得以获得更大的市场份额。与此同时,斯沃琪(Swatch)手表也迅速成为了可收藏物品,消费者选择它们并不是因为手表的精准度或是技术含量,而是因为它们的设计感。相反,有些款式的手表因其独特的设计感反而更加难以读取时间,但同时也更加具有原创性标识。这种对形式的注重超过了对功能性的追求,事实上是对物品形式与功能关系的一种否定,无论是形式还是材料,都没有与最终的使用目的形成一种有机的关系。

产品设计中存在的模糊与错觉,将这一特征表现得淋漓尽致。此时,产品的使用功能已然被隐藏起来,而暧昧的物品造型却得以蓬勃发展。正如,在章鱼的外表下隐藏着一台榨汁机;模糊不清的形式散发出的光芒仿若一盏台灯;我们无法得知手中物件是餐具还是艺术品;我们以为是一支笔而实际上却是一把牙刷。与此同时，有些建筑物好像一个大型产品，而某些产品又好似微型建筑物。

这之间的矛盾十分强烈。事实上，40 年前，作为伟大的译者之一的皮埃尔·保罗·帕索里尼（Pier Paolo Pasolini）就曾警告过我们，这一矛盾不仅仅存在于文学作品中，也存在于农业文明向工业及后工业文明这一特殊的过渡时期。事实上，他已经理解在现代化矛盾愈演愈烈的情况下，"发展"与"进步"的概念在政治理想要求与新兴社会模式中是如何转化的。"进步"是一个不可丢弃的伦理因素，是改善人类生活条件的科学及技术知识的演变。当抽象概念的"进步"与实际动态的"发展"相比时，即实际需求与经济及过剩的生产消费之间的关系。这种矛盾再次出现在"意识"中，它以进步观念支持着人权的意识形态。而以消费意识形态为基础的"存在"，也就是发展的价值。

• • •

技巧与发明之间的设计 > 设计实验的核心是不断地验证理论与实践思维之间的关系。这意味着只有在实践思维存在的前提下，理论思维才会形成。反之，缺乏实践思维，设计就不会在最终的有机体中起实质作用。理论思维来源于我们的文化、认知与记忆。而实践思维则是我们在日常工作中通过使用机械工具来不断地验证、更新技术知识，它是训练与实践的成果。

设计一劳永逸地修正了产品的形式与性能。然而，无论技术的本质是什么，它都能让产品拥有具体的形式，因此，设计总是不断地受制于技术的发展。但是如果缺乏实践思想，设计则是一个脆弱的主题，仿佛艺术制作一般。

通常在设计之前会制定计划，比如，在计划表中写着"前厅"一词，那么建筑师就要考虑如何将它转化为进入房屋的场地。衔接计划与设计之间的空间好像是一份"食谱"，其中列明所有必需的"配方材料"。这一衔接过程在设计之初就已经发生了，但又或发生在设计之时，无论如何，它代表了重要的时刻。在这一时刻，设计师从控制感性设计及理性技术的关系出发，仅靠一张白纸和所需"配料"的清单，加上他的记忆库，以及他对历史，对自然及人工材料，对技术和生产系统，对现实矛盾的了解。此阶段唯一能够确定的是：设计主题、实践思想以及目标对象和对应场所。

从涉及的主题和场所出发，产生的冲突不断发展并贯穿于整个创作过程。冲突涉及形式、制度、人类、材料以及构成体系。在知识的持续累积中，通过相同的构成逻辑寻求到解决冲突的方案。即一种脱离分析但与主题及场所相结合的情感诉求，它随着对主题及场所的了解增长而增长。这个最初的情绪数据与一种特定类型的解决方案相结合，在其功能、结构、形式特性中被分解与分析，然后再重新组合。对构造工艺长期的经验累积与深入的了解，这一阶段显得尤为重要。这个过程往往能够恢复设计实践及空间探索的正确张力，并在艺术维度中重新赋予人工造物、建筑以及技术人员以全新的意义。

马可·维特鲁威（Marco Vitruvio Pollione）对于科林斯柱头形成的叙述，完美地解释了工艺技巧与发明创造之间存在的关系。

§

[1]　设计 _ 文本与语境

马可·维特鲁威（Marco Vitruvio Pollione）写道：
"一个科林斯女孩，在适婚年龄死于疾病。
葬礼之后，她的姨妈将她生前喜爱的
所有杯罐放在一个圆形的篮子里，
并用瓦片覆盖，
她把篮子放在墓碑的顶端。
偶然间，篮子下方长出莨芳花根。
春天来了，花根长出茎和叶子，缠绕在篮子周围，
瓦片边角处受到重力压迫的茎叶，
在末端弯曲呈盘旋状。
当因完美雕塑作品被称为雅典全才的卡利马科斯路过此处，
看见了被茎叶环绕的篮子，
他被这种外形美妙的新奇组合打动，
并以此为原型创造了科林斯柱头，
同时也制定了科林斯式柱型建筑的复杂比例。"
摘自马可·维特鲁威（Marco Vitruvio Pollione）的《建筑十书》第四卷

我想强调的是，卡利马科斯（用技术升华物质）之所以创造了科林斯柱头，正是根据理论与实践相结合的方法。柱式的比例原则综合材质的加工方式，在最终形成的建筑造型上形成了完美的平衡。

● ● ●

材料的真实与虚假 > 在过去的五十年间，建筑物和产品的制造技术发生了翻天覆地的变化。在建筑物逐渐非物质化的同时，机械化、电气化、信息化进一步丰富着建筑本身的性能。这些创新技术实现了建筑外壳与支撑结构的分离，并在构成骨架外形成一层表皮。无论是从地基到设备，还是从结构框架到建筑外壳，在建筑系统与部件的结合中，这些创新技术逐渐改变有机建筑物的统一性及同质性。这些部件慢慢独立，开始有了各自的标准和专业特性。不难想象的是，随着远程信息技术的兴起，这种趋势会进一步发展。同时，这也表明引入与使用时效和形式价值相关的调节系统和通信系统极其重要。

举一个众所周知的例子，即被密集使用的空调。过去，当天气干燥气候炎热时，厚重的墙体隔绝热气，延伸的屋檐遮挡阳光，横向的通风循环空气，我们还可以打开窗户，拉上百叶窗，甚至打开遮阳篷。但是使用空调之后，在一个完全密封的环境中，我们却不能打开任何一扇窗户，即使气温舒适，也让我感

到无助。通常,在这种反常的机制中,使用一种技术来弥补另一种技术的泛滥造成的问题,似乎是正当合理的。那么,可以说空调在密闭环境中的使用是合理的,因为汽车的普遍使用而产生了空气污染。一个闭合的死循环就这样形成了。

在建筑领域发生的情况同样也出现在高科技产品领域中,而且愈发明显。这些高科技产品更加轻盈,拥有更高的性能,它们的重量和性能往往是越来越成反比,比如 iPhone,又或者被苹果的广告词称为"轻于微光(轻于时代,先于时代)"的 MacBook。苹果公司解释:"对于 MacBook,我们给自己设立了一个几乎不能实现的目标,也就是在更为轻盈紧凑的 Mac 笔记本电脑之上,打造全尺寸的使用体验。这就要求每个元素都必须全新构想,不仅要更为纤薄轻巧,还要更加出色。最终苹果带来的,不仅是一部全新的笔记本电脑,更是笔记本电脑的未来。"

苹果公司迷一般的语言,引发了另一种有关材料使用及语言表达之间的矛盾。随着建筑和产品的逐渐非物质化,除了传统材料的阐述方式,还存在另一种让人不安的方式来表达正在发生的转变。我相信,这是更有效的方式。我希望将这段的主题定义为"材料的真实与虚假",这是一种表达材料及其关系的方式。因为在设计中,我们必须考虑到材料与个人情感、历史记忆之间的关系。

材料是人工造物的"文字成分",因为材料让它们更加具化。正是这些材料,构成了人类场所的实体形式。随着时间的推移,材料与生产技术相互作用的同时,塑造并改变这些技术,使其不断地适应各自的特征。这些材料以艺术为前提,因此它们能够直接且强烈地唤起一种文化的连续性。石头、粘土、木材成为制作物品的原料,不仅因为它们是当地的自然资源,还因为它们所具备的性能:稳固性、持久性、可塑性、易操作性、美观性以及与其它材料的融合性。材料作为现实生活中的突出元素,可以回归至我们的技术时代,以全新的形式或全新的组合进行再次构思与诠释,同时与技术交叉,通过自身文化"印记"的伟大力量来为实验与生产指明方向。

无论是在自然环境还是在人工场景中,材料时而真实时而虚假,时而可靠又时而模糊。泥沙除了颜色以外我们什么也看不到,这时它是虚假的;但是同样的泥沙如果向我们诉说一段历史,并为我们的想象力指明方向,又或是向我们讲述赋予它色彩与形式的故事、自然与人类,这时它是真实的;泛着涟漪的水纹告诉我们它的波动是受水流的驱使,这时它是真实的;同样的水用在抽象画中,就好似永恒的壁画,这时它是模糊的;当金属被铁锈腐蚀,表面遍布由侵蚀破损以及光影折射产生的斑点,这时是虚假的;纹路金属板的加强系统使其具有静力效果,并能够垂直放置,这时是真实的;在设计中使用金属铁迷人的几何轮廓,但是却没有传达具体的性能与功能,这时是虚假的;弹簧的螺旋图案却让人联想到特定功能,这时是真实的;借助网格模式的铁栅栏在空间呈现出类似于三石塔的三维体系,这时是真实的;建筑结构中的木头,从它的轮廓和形态可以看到所有制作工艺的历史,这时也是真实的。大地,一切开始的源泉,为想象开辟了空间。当我们穿越自然景观,快速干燥的缝隙看起来像深深的峡谷,又像是被太阳晒干的高原。一面紧凑的石灰岩采石场墙壁,透过其中少数的几个加工工艺,描述了人类改造大自然

[1]　设计 _ 文本与语境

的艰苦劳作。正如根据几何规则放置在剧院的方形石块，能够让自然洼地变为像锡拉丘兹剧院一般的伟大建筑；或者持续不断地将悬崖峭壁变为人类村落中光滑墙壁的基石。

更普遍的是，开采土地获得的原料成为墙壁、拱形系统、衔接空间的制作材料，它们都保留着这种双重特性：真实与虚假。当使用单个石块去制造柱子时，它粗糙却真实；柱头使用的雕刻石块形成了明显的平衡构造，它也是真实的；在高迪的桂尔公园中，人工技巧仿佛虚幻，而大自然展示了真正的造物空间，它依旧是真实的。帕埃斯图姆神庙的每一根石柱，用它剖面的凹槽向我们描绘了精致的雕刻与熟练的技艺，通过明暗对比给人一种捕捉住光线的感觉，这时它是真实的；最基础的三石塔结构，两根柱子一根横梁变身为建筑体系中的一件艺术作品，这时也是真实的。相反，在哈德良离宫中的海上剧场，矗立着几何形状的石柱，在它们光滑表面的背后隐藏了艰巨的工作，这时材料就是模糊不清的。

材料的真实性体现在生产制造过程中，具有具体的项目、模块化和系列化准则。例如，砖块作为最小单位的发明使其应用于两大领域：需要时间的现实合理生产中；由砖块的重量与尺寸决定砖砌结构内在规律的灵活设计中。

透过砖墙和石墙，人们可以看到最终形式背后的所有工序。与它们不同的是，高科技幕墙所使用的材料则完全非自然化。它们的屋檐正逐渐消失，金属纹理越来越纤细光滑，表面所有物件被去除，随着外形、材质与色彩的"渐变"，其拐角被磨平，门窗与墙面平齐，建筑物的立面呈现为单一的玻璃表面。

一系列规模上的飞越，作为基础配件的玻璃板成为建筑物的最终参考。此时，建筑物被视为科技物品的范式，借助其中的精细技术它展示了具有复杂性能的基础部件，这些部件构成及隐藏了建筑物的承重结构。通过一系列特性的转移，这一现象同样适用于产品领域，对于产品而言，构成材料的真实性和虚假性成为决定产品卓越与中庸的重要因素。

有关近未来设计 > 在不久的将来，设计方式会发生改变吗？很显然，与过去相比，设计生产产品的方式在长期实践中表现出了强烈的不连续性。这一现象的产生，归根结底是因为生产环境的改变。我们将其定义为后福特主义，它处于后现代主义文化当中。后福特主义与后现代主义这一复杂的组合概念，通过难以避免的技术发展，带来了开放源代码（open-source）文化。新的信息科学产生了新的设计现象，以3D打印机为例，它并不是什么特别复杂的物品，却能够实时地将想法转化为现实形态。这样的方式打破了产品物理结构的传统顺序，从设计生产到市场营销，改变了垂直产业链中涉及的不同专业人员。现在所存在的产业链则是扁平化的，在需求数量上以定制化为主，因此不受大规模生产的束缚。但值得注意的是，"在这种情况下，设计应当更具影响力。也许设计方式会发生改变，也许产业链分工会有所不同。但是，至少设计师可以重新进行思考，发明新的产品类型。另一方面，如果我们变身设计师，每人都拥有一台家用打印机来进行设计生产。那么，造成的风险可能并不是无穷无尽的变化，而是产品类型的激进化，这最终会形成我们物质文化的典型印象。"[6]

设计人工造物及相关主题

我感觉到一阵风，正持续不断地吹向创新需求的相反方向，也就是需要摆脱后现代主义文化的有害影响。后现代文化所处的社会是流动的，它试图提出新型民主参与的有益模式，但由于文化限制的缺失，最终由"小叙述"＊替代"大叙述"。

上述"信息社会"，"后福特主义"和"后现代主义"预示着社会上传播所谓的民主福利带来的有益效果。然而在我看来，它们似乎背叛了他们的理论使命，这是致命的。相反，在现实社会当中，却呈现出了独特创新的泰勒制延伸，因此，在不断升级的社会混乱中，使工作时间与生活时间冲突不断。

事实上，我们可以对这一矛盾进行双重解读：根植于本土的小型企业的普及，它们依然以雇主与劳工之间的人际关系为特征；大型企业试图打破劳工市场规则，摆脱工会控制。但是新的地方保护政策真的能够减缓三十年的发展趋势，促使跨国工厂返回本土，把工作机会从中国、墨西哥或印度撤回吗？苹果公司作为高科技领头企业，拥有以成本计算和质量为基础的生产物流链。他们在日本、中国台湾、德国生产复杂的零部件，最终在中国组装成品。将众多子公司和供应商设置在同一个地方，同一个国家，这是一个漫长且昂贵的操作，这也是不可能发生的。

当今的资本主义应当灵活面对截然不同且多样化的消费者需求，因此，生产领域应更加灵活。即通过新的信息技术和适当的劳动模式辅助，即使是面对不同的小型的消费群体，生产者也应当具有根据需求波动改变生产的能力。根据后福特主义理论，这种有利于小型企业发展的情况，进一步产生了资本的混乱，除了社会关系以外，还影响着商品的生产。

在这种不稳定的环境当中，设计的技术支持与设备将会发生改变。产品本身与产品支持以不同的方式来满足我们尚且无法想象的需求。但是我希望，设计师始终能够将人类的情感（情感于技术之前，同时对美学进行完善）作为衡量需求的核心。事实上，这些情感是设计伦理的来源，设计师肩负着道德责任，并不断地改善设计作品的构思概念，试图将设计伦理注入其中。我相信，即使是再先进的技术，只有成为人类思想快速发展的有用工具才能最终实现其价值。

[1] Umberto Eco, *La società liquida*, in:"Pape Satàn Aleppe"; La nave di Teseo, Mi-lano 2016.

[2] Andrea Branzi, *Neo-prehistory: 100 verbi*, in:"Neo-prehistory: 100 verbi," Triennale di Milano and Lars Muller Publishers, Milano 2016.

[3] Kenya Hara, *Poesia sul desiderio degli essere umano*, in:"Neo-prehistory: 100 verbi," Op. cit.

[4] Raimonda Riccini, *Il senso del design per il tipo*, in:"Type & Model | idee, progetti, azioni," Planning design technology Journal n. 4, Rdesignpress, 2015.

[5] Roland Barthes, *Giocattoli*, in: Miti d'oggi, trad. Lidia Lonzi, Einaudi, Torino 1975.

[6] Raimonda Riccini, *Il senso del design per il tipo*, Op. cit.

＊ 小叙述：着眼当下的哲学思想

§
[2]

文本与语境

§
[2]
p.27

引言

> 工业设计与设计
> 构思消费品
> 事物、物品与产品
> 伦理与美学
> 有用与无用
> 混合现实
> 文化创新
> 传统与变革

回顾近几年我接触的与设计理论相关的作品，让我意识到这其中很大的一部分是文集，是以一种收集评论文章、书目内页、作品，设计师、作家的方式呈现。我还认识到文集的形式具备它固有的科学及学术用途，体现出摘录者选择他人文章的责任感，还试图找到一个点来平衡原作者的语句和摘录文字所表达的意图。这种情况下，即使被研究的对象也是执行者本身，它依旧显得具有欺骗性，无论如何，其中的不确定性是不可避免的，因此也难以时时保持自我的评论意见。正如之前说的一样，为了便于理解，这本文集利用阅读关键词的形式在教学框架内衔接摘录的段落。对于所选需要验证的对象，我采用拆分重组的方法，我希望这样的叙述方式更具说服力，能更好地通过时间轴以及个人设计价值理念进行表达，虽然我的设计观念仍有待验证。

§
[2]
p.28

工业设计与设计

无处不在
摘自《设计无处不在》，2013 年 1 月讲座

设计自主
摘自《真实与虚假的辩论》，2012 年《DIID》第 57 期

不可避免的工业化
摘自《工业的自然性》，2012 年《DIID》第 55 期

教授设计
摘自《在相同与差异间的设计》，2009 年《DIID》第 42—43 期

物质与材料
摘自《材料与自然》，2008 年《DIID》第 38 期

教育与培养
摘自《学院之后的设计师》，2007 年《DIID》第 32 期

相同与差异
摘自《不同与设计》，2006 年《DIID》第 24—25 期

展示活动
摘自《设计展会》，2005 年《DIID》第 17 期

复杂的虚拟结构
摘自《图形及多视觉多媒体通信界限》，2005 年《DIID》第 16 期

它是设计
摘自《设计无处不在》，2003 年《DIID》第 7 期

物品的语言
摘自《全球化设计／附加道德》，2003 年《DIID》第 3—4 期

未来场景
摘自《工业设计》，2002 年《DIID》第 1 期

很多情况下，我们十分谨慎地将"设计"或"工业设计"分开进行表达，这有时是有利的，有时是多余的。相对而言，前者更灵活，更生动，更无时间感；而后者更明确、更清晰、更具历史感。如果说"工业设计"代表了对具有现代性产品设计形式的解放，那么"设计"则成为一种表达系统，代表了在单个产品之前所衍生出的所有需求及其使用条件。这并不是在两种立场间进行选择，而是叙述方式的不同，在叙述设计故事的时候，我们应当充分意识到该从哪个词汇开始。

无处不在 ＞ 设计无处不在，因为它改变着外观，它像一只变色龙。设计无处不在，因为它打破了所有对后工业的预测，每天创造新的事物来表现新的形式。设计无处不在同时还体现在伦理和援救方面，以及商品学和形式主义的诞生方面。

设计无处不在。源自工业设计的实践文化，意识形态文化及手工传统文化，推动着设计的形成。

设计无处不在。它越过经济、工业主义文化、现代文化以及后工业主义文化，直至非物质文化和信息社会、经济时代的到来。

若设计是明确的，设计则无处不在。罗马的院校给出相应的诠释：人造的产品，所有都是经过人工而成的；设计表现了物质文化；设计阐述了人类总是将新的需求变成目标产品来改善自身的生活和社会关系。如果我们考虑到现实生活中的物品使用的次数及其作用，那么，设计的无处不在就显而易见了。但是这又表明我们只能进行表面预测，因为我们周围遍布的大量物品会发生难以置信的变化，尤其是理论的变化。这说明了什么？设计已然成为了全球性文化，以创新发展的名义，设计穿过全球而没有必然地产生稳定性，更确切地说，在经济、社会与文化之间的不平衡或"动态平衡"，设计是不能解决问题的，但可以持续地提出新的问题。"消耗品"有着难以想象的价值，但这不是对这种模糊概念进行单纯的理论研究。然而事实上，设计的运转使得动态平衡系统永远无法停止，为解放"提案文化"先于产品创造了条件。只有以这样的方式，设计才能越过任何学术与科学争议，赋予合乎情理的必要的培训以意义。

在设计师养成方面，无处不在的设计表现出了一个不可辩驳的事实。近来，专业设计领域学科秉持着遵循已久且一直遵循的原则，反思了制造领域的变化，并提出了不同的解决方案，接受了培养设计师形象及其发展的艰难任务。这任务并不简单，同时也伴随着众多的不确定性。

设计无处不在，不是因为到处都是物品。无处不在指的是在大量物品之前进行的大量设计方案。Quaroniana 推广设计能力理论的提出，让预测现实世界的微小变化变得可能。因此，设计通过微小的革新促进改变。没有革命，没有宣言，但是保卫着任何地方都有可能做设计的原则。总之，物品并不能证明理论，但有可能对其进行阐述、反对及验证。但是要注意，所有的这些是人人都懂的，所以似乎是民主的，也无法避免的。无论如何，设计要为潜伏的民众提出的方案承担部分责任。

无处不在的设计发起不同的口号，从"设计所有，直到为所有人设计"开始，为它的学科及文化扩张打上烙印。设计无处不在是因为它自发地发生改变。不再如同七八十年代，而是提出一个正式的能解决建设性问题的方案。如今的思维孕育着经济改革，设计有可能成为改革中重要的能量提供者，也就是说，"涉及所有的全球产业的巨大需求"。在这些前提下，安德列亚·布兰茨（Andrea Branzi）认为，设计不再是用来生产杰作，而是用来改变场景，从而引起意识的改变，因此需要持续不断地提出想法。当对更高水平的技术创新进行探索的时候，设计无处不在。此时，通过制定出针对前景生产的战略，设计还没有投入生产的科技产品，预测新的难以想象的人造物性能、新的材料、意想不到的应用，统一市场来满足发达工业社会和发展中社会的需求。

设计自主 > 自文艺复兴以来（也许更早），产品的设计与生产正逐步分离，涉及人员的角色与职能也最终得以区分。如今，许多手工艺人具备"动手能力"，但是他们并不需要同时具备设计和生产的能力，来创造出那些过去没有的东西。他们可以进一步完善巩固某种产品类型，而不是去取代它们。随着达尔文进化论的提出，科学发现的进步，从棍棒、石头、长矛、弓箭、石弓，到手枪、步枪、机枪等，这些器具类型的不断演变，慢慢地将新事物的设计者和生产者分离开来。在工业革命的过渡时期，出现了大规模的工业生产，设计与生产实现彻底分离。此时，设计师的形象已变得完整，设计也成为了一门独立的学科。这个结论正好与沃尔特·格罗皮乌斯（Walter Groupius）在创办包豪斯（Bauhaus）学院时（成为横向跨越艺术、设计与建筑的不可复制的重要时期）的想法不谋而合，他也认为应当进行明确分工，并为此付诸努力。包豪斯设立之初，设计师们提出支持这一分离的宣言，也绝非偶然。密斯·凡·德·罗（Mies Van der Rohe），保罗·克利（Paul Klee），康定斯基（Kandiskij），汉斯·迈耶（Hannes Meyer），他们是二十世纪在产品技术与文化之间博弈的新一代关键人物，他们设计的许多产品至今依旧风靡于世。但这也并不是说约翰·拉斯金（Jhon Ruskin）和威廉·莫里斯（William Morris）主导的工艺美术运动没那么重要。那么，还会有哪所学校不开设自主设计课程？这值得深思。尤其是设计混合产品时，是否存在一体的自主设计与生产，我想，我需要谨慎思考。正如米凯莱·德·路奇（Michele De Lucchi）在他的《私人收藏》中提到：这是一种从手工业文化中得到启发的超工业实验机会，它不仅是一种手工劳作工具，还是一种智能手段。也就是说，发现与使用手工艺，通过不断地渗透，也能"同时地"运用体力与脑力，"同时地"结合传统与创新。

不可避免的工业化 > 从纽扣到磁卡，这些自身作为原型的"事物"，赋予了设计一种合乎情理的形式，即一种简单而又大众的解决方案。在此，观察力、实用性以及美观，它们相互影响并相互制约。为了真正达到优良设计，在设计与工业生产过程中，产品背后的知识专利与设计师署名并没那么重要。
重新思考这些"事物"在设计生产过程中背后的"作者"，这并不是偶然，也不是现在才开始的。在后工业体系中，广泛传播及分享设计的作用之一就是描述开放源代码（open-source）这一复杂的设计现象。制造商、自制者、社会设计，有很多方法来确定现代设计的联合价值。特别是当需要通过政治文化价值来充当在"无名才华"与"实用美观的新范式"之间的桥梁。如今，这些范式在任何形式状态下，也都被大家接受。
这些想法让我们回到40年前。1972年，布鲁诺·莫纳（Bruno Munari）提出与其将"金圆规奖"授予设计师，不如授予物品。在他看来，这些物品虽然脱离时尚，但是它们在材料、技术与功能中达成了完美的平衡。因此，这些物品没有任何多余部分，且恢复了从设计到仓库的经济学思维。虽然有关授奖，还需要讨论造型美学，但是在"设计问题"的内部逻辑中就能找到该争议的解决方案。对布鲁诺·莫纳（Bruno Munari）而言，这些物品的本质，是不畏惧创新技术和设计语言的改变，即使将来诞生更新的材料和技术，它们也会从中发现并利用，借此延长物品的寿命。同时，这些物品从不表现模糊的符号，更不属于特定的社会阶层，它们让智慧的民主和高品位的主权变得切实可行。而智慧与品位，随着历史的演变，依旧保留着这种不可避免的工业化的自主形式。

教授设计 > 在工业设计概念中，设计师是日常生活中产品流的产物。他们反应出在创造方面，需要通过大量的学习实践来认识、理解和学会制作：即需要学校。所以，想要了解如何在欧洲大陆上表达创意，就必须确定设计表现的基本特征。看看刚从学校走出来的年轻一代，听听他们在专业领域的第一次经历，也就是了解"学院之后的设计"（Design after school）。无论是从斯堪的纳维亚学校到盎格鲁撒克逊学校，还是从欧洲中部到地中海区域（意大利、法国、西班牙、葡萄牙），上世纪的历史经验和学院特色可以作为欧洲设计一致的参考。

19世纪到20世纪之间，随着工业革命的爆发，设计的诞生，盎格鲁撒克逊学校的基本特征发生了根本性的改变。在英国，工业革命引发了工艺美术运动之类的运动。在工业生产和手工生产的对立交替中，无论是美学还是意识形态的原因，工业领域的分离性质与手工艺行业的创意性质并存，直至亨利·科尔（Henry Cole）时期，才明确提出需要为新时代培养专业人才。在开设的学校里，通过教授新的材料、性能和美学，包括生产方式和技术，来打造和输送设计人才，以此营造工业生产的环境。除了学校以外，还开设了应用艺术博物馆，向大众提供了竞争对比的机会，打开了类似于全球展会的大门，以此推动工业发展。在这样的环境背景中，诞生了由威廉·莫里斯（William Morris）开展的工艺美术运动，以及皇家艺术学院（Royal College）这样的院校。在此后的1967年，皇家艺术学院的目标宪章（Royal Charter）中提到："在艺术与设计领域，通过进一步学习，掌握特殊的专业知识与技能，包括工业程序、商业贸易、社会发展以及其它与工业和贸易相关的教学、调研与合作。"从辩证的视角可以看出具化的实用主义与不确定的创造能动性、对产品的技术控制与实验模糊性的绝对张力、对产品性能的控制与技术创新。皇家艺术学院这一方针，至今为止依旧代表了盎格鲁撒克逊学校的教学特色。

前有德意志制造联盟（Werkbund），后有包豪斯学院（Bauhuas），这些机构的经验表明，现代主义运动的方针与理论显然已经渗透到欧洲及欧洲以外地区，以致他们的方法及表现形式被历史学家定义为国际化风格。

物质与材料 > 最初，物质是由"自然形成"的方式塑造而成，也就是思考什么是真正有用的资源，来直接生产人类需要的东西。石头、粘土、木材之所以被选为建筑房屋与构造物品的材料，不仅因为它们是可用的自然资源，还因为它们本身的构造能力，它们具备的静力学、持久性、可操作性和与其它材料结合的能力，以及它们对形式与美观的不断追求，它们甚至试图代表一种绝对的价值体现。更确切地说，我们注重所谓的天然材料，与其说因为它们的可用性，不如说是功能性。

但是，随着时间的推移，我们对材料的思考方式发生了深刻的改变。一方面，通过手工艺操作使材料产生"反应"，通过不断地塑造材料的外形来不断地适应新的环境。另一方面，利用创新技术改变材料的形式，赋予材料更多的可能性。因此，天然材料和人工材料越来越难以区分。事实上，21世纪确定了一种新的理解材料的方式，即材料不仅是产品的构成成分，也成为其外观与功能的一部分。随着这种理解方式的诞生，我们与日常物品之间关系的物理参照和语言，发生了翻天覆地的变化，这些参照和语言让我们通过所有感官与材料进行交互作用。正因如此，改变我们与物质的关系，一方面打开且激发了空前的情感压力，增加了设计的可能性，另一方面，无论是有关材料特征，还是有关材料能做什么，之前的平衡与参照都被一一打破。

过去，为了在新产品中接纳新材料，模仿自然可以作为一个理论模式来达成目标。举例说明，当塑料（人造树脂）最初被采用时，用一种自然的形式将所有产品表面覆盖一层仿真表皮。这样一来，这些产品可以使用未知的材料进行制作，绝对实用且成本低廉。它们使用不同种类与形式的仿制木纹进行装饰，尽可能地模仿"自然"材质在记忆中本身的样子。这一举措隐藏了材料的本质，覆盖上我们熟悉的特征，试图唤醒那些众所周知却无法言说的感受。可以说，最终我们看重的并不是材料的真实性，而是材料试图传达的不变形象。

教育与培养 > 近年来，意大利大学体系在设计人才培养方面，提出了高要求的任务，同时也热衷于加入由不同地区形成的国家教育网络来分享不同的经验。虽然各个学校教育的基础方法和理论有所不同，但是大家都有着共同的目标：培养年轻的设计师，让他们得以在必要的时候进行学习，适当的时候进行领悟，以创新来竞争，以想象来成长。

私立学校作为重要的教学机构，只有在极少的情况下会搭建年轻设计师与企业间的专业关系。一般来说，企业与学校之间的关系是以双赢为基础的，为企业开展新类型产品调研的同时，学校会以此为题设置对应课程。这样一来，企业既不需要制作模型也不需要将概念工程化，就可以以最低成本获得"灵感仓库"。公立大学教育的不同之处，在于与企业的关系主要以达成实习合作作为主，让学生能够跟企业进行更直接的接触。通过实习导师的帮助，学校持续地将年轻有为的学生输送进职场，并保护他们在专业领域的"创造力"。显而易见，年轻设计师为企业设计的产品数量不能够代表全部，但对于我们寻找的答案而言却意义深远。与此同时，在选择年轻设计师的作品时，无法避免会存在一些个人倾向，不能做到完全公正。想想在米兰三年展博物馆举办的《新—意大利—设计》展（2007年1月20日至2007年4月25日）。该展览由希尔维亚·阿尼卡利克（Silvia Annicchiarico）构思协调，安德列亚·布兰茨（Andrea Branzi）策划完成。事实上，这次展览就相关文化背景及其选择产生了一系列的争论。安德列亚·布兰茨（Andrea Branzi）通过他选择的产品和项目来证明他提出的论点，即意大利是创意工作室的聚集地，这些工作室传达了一系列有价值的新思想，它们传播了创新文化，形成了意大利设计背景。

工业设计与设计

相同与差异 > 在设计领域，"相同"与"差异"有着不同的作用，但是在全球化市场中对工业制品进行的判断却是一致的。在工业设计领域的某些类别中，对"差异"的定义值得进行详细透彻的描述。其中包括属性的差异，是指根据目标对象（女人、男人等）来区分相同用途的产品；种类的差异，是指即使这些产品遍及全球，不同的物质文化具有不同的表达方式，它们却依然保持与目标地域环境一致的特征。这让我想到了许多烹饪与饮食的器具；物品符号价值的差异，是指在一组功能相似的产品中，符号价值已经成为产品的一种竞争力。在这样的情况下，产品的自身特征比技术性能的质量和数量更为重要（iPod最具有代表性）。当然，更有趣的是分析如何构建"相同"与"差异"的概念。工业设计涉及了遍布市场的各种各样的产品，它们针对不同的历史特征和不同的文化背景，面对着具有不同需求的社会团体。这就是为什么工业生产都是大规模进行，并不时地试图定义一个标准，也就是说，这就是预定义一个特征或一组特征的原因。因为通过制定的标准，可以将产品中的"多样性"升华为一个"特定特征"，来满足全球市场，即标准化使它们符合多样化的需求。

展示活动 > 国际设计巡回展是一个阐述产品及其定位的关键环节。展览上不仅展示了产品本身，同时，也直接或间接地体现了企业与设计师的技术与能力以及其文化背景。

展览是为了更好地推广企业及其产品，因此，橱窗也会定期更新展示的产品种类及对应的设计语言。事实上，消费品被广泛地使用在各种生活及办公场所，并应用于各种日常活动中，而正是通过这些展览开创了它们的流行新趋势。通常来说，展览是以商业性为目的，但是不排除它们试图落实实验操作，挑战创新前沿的可能性。同时，展览也能够为设计师和企业赢得声誉，为新产品投入市场创造机会。最后，展览也提供了可以进行交流、专业出版和展厅设计的机会。越来越多的媒体开始关注展示活动，密集的展览日程吸引着广大群众，同时也带来了更多的商务机会，大量资源的投入使得展览越来越惊艳。在这种情况下，若想将产品面向大众，首选的战略思路是将其置于精心设计的故事场景之中。如今的产品设计展也正是如此，更加注重媒体传播和展览的创意形式。慢慢地，参展企业也更加重视产品的创新生产过程、类型及其性能。他们对展厅设计的追求已远远超过展示产品自身。

早在2003年4月Fuori Salone*的《人间天堂》（Earthly Paradise）展中，亚历山德罗·门迪尼（Alessandro Mendini）为展示活动制定了总体计划，他阐述了对应主题，定义了总体规则，让一些权威设计师以此为题对展览内容进行设计。《人间天堂》的主题为设计形成一个空间，在这里，人们花时间投入自我，投入身体和情感，在这里，每个感官都得到释放——这里就是人间天堂。

同时，设计展览也可以成为非物质表现的机会。同年的三年展博物馆中，妹岛和世（Kazuyo Sejima）和石上纯也（Junya Ishigami）为雷克萨斯品牌展览设计了一个混沌的空间。他们让汽车展览直接变为一

---

* Fuori Salone，米兰设计周会外展

个乳白色消逝空间，其中，所有的实体物质都因为缺乏色彩而消失于雾气当中。通常，展览设计可以让很多产品受到大众的追捧。同时，展览设计也可以探索具有绝对价值的场景效果。2005年的 Fuori Salone，威达（Vitra）公司展出了布鲁莱克兄弟（Ronan & Erwan Bouroullec）的作品，但这个展厅设计旨在促使观众将焦点放在前景处，而展示出的系列产品却几乎失焦而沦为背景。

此外，展览设计还可以以追求惊艳效果为终极目的。正如米兰城区的一只巨大的毛毛虫，充气结构让它灵活且便于移动，特殊的外形与尺寸使它成为了极具交流能力的符号，它趴在建筑物的屋顶上，用它的大脑袋面向来来往往的观众。它有着简单且迷人的结构，白天清晰可见，夜晚闪闪发亮，人们无一不被吸引着进入其中一探究竟。在毛毛虫的内部空间中，可以进行会议、主题展览，也可以是企业、设计师和相关机构的展示场所。

展览设计仿佛是当代社会缩影，比如在东京的设计周或100%设计展览中，这些展示空间被布置在集装箱中，长达一年。这些集装箱的安装与摆放，构成了一道靓丽的城市风景线，好像汇集了全球仓库贸易的商品一般。集装箱的布局试图构建并成为《城市展览》的一部分，《城市展览》中的各个元素形成了真正的展示设计目录，总结了展示设计文化的所有可能性，同时也更加突显了设计师这一角色的重要性。

复杂的虚拟结构 > 信息技术在设计中的广泛应用让平面设计师的工作发生了根本性的改变。过去通过不同的语言以及平面载体，以手动操作的图文印刷形式来表达及沟通交流。而如今，视觉传达设计师已然涉足了更多的专业领域。平面设计不再局限于以画报书刊为载体的单一表现形式，而是演变成以网站为主的三维的、动态的、交互的新形式。

过去的平面通讯系统依赖传统的概念，而如今，这些通讯系统可以改变字体、图形排版，并借助视频的使用。如今网站的复杂虚拟结构是设计师视觉传达最重要的表现领域。以场景布置为例，其内部语言受到许多制约，如使用技术的网络循环过程，或是实现用户与其进行交互的软件与工具，又或者是形成不同信息级别的透明层系统。

网络世界充斥着由不同文化构成的设计图形，因此，设计师必须认识到由网络引起的零距离的沟通趋势，并提出一种全球化交流语言，以此建立共同的财富。

它是设计 > 设计与工业生产，产生了毫无限制且用途广泛的消费品，让我们被有形或无形的产品包围着。因此，设计可以以其不同的核心内容来改变社会风俗，影响人与人之间的行为。同时，在产品生产与创新过程中，设计决定着物质对环境产生怎样的影响。

在这样的氛围中，同样的设计概念，不再是简单的专业术语，也不再是现实生活中的产品制作。而是重新定义的一种意识活动，一种对满足新旧需求的物品和（或者）服务的反馈能力。

当代社会的设计，与其说是为精英文化服务，不如说是为日常生活设计。也就是说，设计适用于广大消费品，它针对所有生产和销售部门，用清晰或隐晦的方式表达设计需求。

设计新的产品，是在这些产品的生命周期结束时，延长它们的存在。借助其零部件的再次使用或者对材料及零件进行回收处理，让它们以其它形式存在。

设计新的产品，是思考真正可替换零件的系统，通过不同的应用与组装方式，让更多不同的模型得以呈现新的形式。

设计新的产品，是首先实现其基本功能，然后通过增加辅助功能，来丰富产品复杂的功能性。营销策略的设计，即设计与商业化和工业产品消费相关的无形服务，以确定除了达到服务的自身价值之外还能带来盈利。设计师是一个非常有趣的职业。通过个人学习的积累及所处的社会背景环境，设计师可以通过分析环境和文化因素的影响，来设计引发购买欲望的特定产品的广告海报。设计，逐渐成为一种将社会中观察到的数据转化为综合表现的工具，正如手机产品是最重要的应用领域。因此，设计师也是一个非常复杂的职业，他们通过一段又一段的经历来充实技能，让形象更加专业，同时借助信息自动化技术的不断发展来更好地组织生产和消费。

设计，其传统的任务是赋予产品外形，如今，它涉及的是整个复杂的工业领域。因此，设计师的养成需要长期且系统的学习。他们需要接纳先进的知识，理解供需关系；认识并了解应对复杂性的工具与方式；不论是在性能方面还是在生产技术方面，提供工具来分析需求，并将其转化为创新产品。

物品的语言 > 工业产品在它的初始阶段，就呈现了伦理道德与全球文化。设计方案将概念深入为提案，然后确定其制图、模具、模型、造型及其功能，并为用户提供产品的使用说明。当用户抓住产品时，通过设计方案来确定产品对于触觉、视觉、听觉，甚至嗅觉和味觉的敏感点。设计方案形成了集零件、机械部件、电路于一体的复杂系统并确定生产方式及所需时间。毫无疑问，设计方案确定了物品的语言、风格及其自身的附加价值。产品的设计语言已然成为其中的一部分，且持续不断地探索新的特征。产品的外在形象如同碎片，它们既无法击破稳固的系统，也无法超越先驱的霸权意志。产品的风格表现了认同文化的状态，所谓认同文化，是基于对全球风格怪异与迷人的组合的不断研究。无处不在的产品设计的语言，在全球空间范围内，描绘着我们日常生活的场景。

更为明显的是，在生产系统方面，它对传统工艺的操作系统及逻辑进行了革新。可以肯定的是，技术革新确实对这一环节有一定的影响，因此，其中涉及的伦理、社会与经济是极其重要却也是极其矛盾的。慢慢地，产品开始由分布广泛的产业链共同完成，其中安装和储存甚至都不在同一个地理位置。比如十年前，在通用汽车的销售价格估值中，中国台湾、新加坡和日本负责小的零部件，占每辆汽车总价的40%；韩国负责加工和装配的任务，占总价的30%；日本负责生产高科技元件，占总价的17.50%；德国设计外观及机械部件，占总价的0.75%；英国投入广告和提供市场服务，占总价的0.25%；而爱尔兰和巴巴多斯岛约占0.5%，用于执行计算机数据演算。整个汽车生产制造在众多国家地区分布开展，另一方面，这样的布局会限制主要部门的卓越能力和贫穷地区工人的技术能力之间的融合。经济学家、哲学家阿马蒂亚·森（Amartya Sen）在《全球化与自由》一书中指出，不平等现象是地球经济秩序多变的主要根源："虽然我们现在比以往任何时候都更富裕，但是我们所处的世界依旧存在极度贫穷的地区，现实中充满着不平等。"

未来场景 > 每天，我们用双眼（或用框架、隐形眼镜之类的衍生物），观看成千上万的画面。这些画面出现在报纸上，家里的玻璃窗外，移动的汽车外，以及闪烁的屏幕上。树脂、胶水、塑料，空气的味道，不断地刺激着我们的嗅觉，成千上万的新的粒子改变了四季的香气，混合成了遥远大陆的味道。在产品与不同环境之间，存在着一种混合物，它与产品的外在形式、设计语言有关；它与物质世界的建立过程有关；它与城市、人为改造与否的环境关系有关；同时，它也与我们的身体、我们的思想和我们的感官有关。在这个混合物存在的场景中，设计并生产产品，控制它们的生命周期，直至它们消失，这些无不代表了产品的各个综合阶段，展示了多样性的科学技能、创新的艺术表现形式、新的工艺和新的社会伦理相关的复杂决策。那么，对工业产品进行的分析帮助我们更好地了解我们所处的时代，因为，产品反映了所谓地球村的物质文化。同时，也帮助我们更清晰地认识到工业产品设计师是一个复杂的职业，因为他们不再是简单地考虑产品的外观美学，还需要掌握更加复杂且专业的主题观点。如今，设计师处于高层次创新技术领域，他们开始提出以"未来场景"为目标的生产战略，设计未来使用的科技产品，预测新的难以想象的产品性能，以及应用新的材料，来统一市场上人们的需求。

§
[2]
p.37

构思消费品

新型科技
摘自《设计无处不在》，2013年1月讲座

色彩
摘自《色彩讲述所有设计》，2012年《DIID》第53期

教学方法
摘自《在相通与差异间的设计》，2009年《DIID》第42—43期

可持续技术
摘自《生态设计与新场景》，2009年《DIID》第41期

连接
摘自《附属科技》，2009年《DIID》第39期

含糊的概念
摘自《材料与自然》，2008年《DIID》第38期

教育经历
摘自《学院之后的设计师》，2007年《DIID》第32期

市场需求
摘自《设计界限》，2006年《DIID》第26期

惊艳亮相
摘自《设计的惊人效果》，2006年《DIID》第18期

沟通交流
摘自《图形及多视觉多媒体通信界限》，2005年《DIID》第16期

错觉感
摘自《污染、传递与同质》，2005年《DIID》第14期

有关材料的思考
摘自《材料的真实与虚假》，2005年《DIID》第13期

磁卡
摘自《设计无处不在》，2003年《DIID》第7期

有益的陈旧
摘自《感官转移》，2003年《DIID》第5期

俏皮亮相
摘自《玩具工具》，2002年《DIID》第1期

一般情况下我们很难关注到简单的事物。众所周知，当我们对一个现象进行全局观察时，我们常常会忽略其细节。这也就意味着，设计系统的风险在于表现自主与包容的同时，如何运用广泛的设计知识，尤其是在过于关注完整的设计系统时，通常会对设计过程之前、之中、之后的选择判断产生错误。基于设计现象的前提和诠释，它的最终任务是设计人工造物让其行使社会职责，也就是阐述人类的需求及表达其物质文化。

新型科技 > 少即是多（Less is more）的原则，与最初相比，如今呈现出了更多的含义。通过少量使用材料，减少能源消耗，更加注重产品的环保概念，试图在自然与人工当中寻找新的平衡；在加工工艺的微小误差方面，在去除冗余形象和无趣形式方面，它变得更有控制力；它促进了精细技术产品、小型产品和无形产品的设计与生产，利用微处理器技术让产品功能更具智慧性，仅仅通过人性化的手势，不需要很高的文化程度也可以使用这些产品，使得人机交互更加便捷；它最终也成为一种研究方向，在未来场景中运用纳米技术，精简物品本体。创新技术促进了科技产品生机勃勃的发展，迄今为止，这些科技产品作为身体的延伸物，以一种更精确更有效的方式执行着具体功能。其中包括执行人类的感官功能、满足人类自然需求、表达人类与自然的关系、帮助人类建立社会关系，以及辅助人类参与的所有日常活动。

新型技术参与生物加工，创造转基因生物，从而将人造及工业产品的概念延伸到食品行业。比如西红柿、西葫芦等，过去只能任由其自然生长，而如今可以在生长周期中改变它们的颜色和大小来满足不同的市场需求。新型技术让产品更加智能，它们可以与人类互动交流、传递情感、远距离倾听、增加幸福感，使人与人之间的交流更加便捷。新型技术不仅可以将产品用于身体外部，还能将新的功能直接植入我们的身体内部。比如让我们得以在黑暗中观察，在水中诉说与聆听，让我们的视觉超越了真实的维度。因此重新思考人类自身的状态以及和未知难测的科学的关系，显得尤为重要。技术创新具有的强大潜力，也正推动着我们超越所有想象。

色彩 > 在工业产品设计中，色彩的使用，色彩宽泛的主题，色彩的多变性，色彩丰富的文学指向，逐渐让人感到措手不及。过剩的信息似乎并不能给研究人员和学者带来什么帮助。为了克服这个困难，我们选择在可触与不可触的产品中，反思色彩的异质性。在此，我个人发现两个特别有趣的特点。

第一，色彩是设计师个人经验的重要部分。亨利·马蒂斯（Henri Matisse）认为，色彩已然超越他的画作，成为了一种自由。也就是说，通过多变的色彩刺激产生了众多的感受。因此，当我们重新设计色彩符号与装饰时，我们不仅是用不同种类的色彩进行美化，也是在操作由场所、物品及人类之间的关系产生的灵活特性。而这种关系可以通过通感大小来衡量，他们以可见或不可见的形式，影响感官之间的相互作用。这一关系也使得色彩集合表达出一种隐喻及语言方面的复杂性。随着时间的流逝，它们因为不同的文化法则而有所不同。埃托·索特萨斯（Ettore Sottsass）认为，色彩即文字。回顾埃托·索特萨斯（Ettore Sottsass）的经历，我们想象一下，如果可以利用色彩书写"视觉著作"，那么，那些过去与现在的色彩作品就能够跟我们进行更深入的交流。"视觉著作"对于亚历山德罗·门迪尼（Alessandro Mendini）来说，也是一个及其重要的主题。位于最外层的色彩，作为一个书写符号，让色彩表层更具描述性、交流性，也更加感性。

第二个特点似乎显而易见，尽管如此，我依旧想要进行阐述。利用色彩进行设计，同时，色彩本身也是一个设计作品。色彩是一件工具，是一种手段，通过它来实现视觉物品的复杂性。所谓视觉物品，就是遍布周围的物品的色调、符号以及表面与体积间的可见抽象化。通过对原色的对比使用，以及对几何图形的研究，来平衡产品中的功能比例，包豪斯学院已经将色彩主题作为一门独立课程。与此同时，意大利设计也给出了色彩最全面的展示。无论是使用夸张的手法还是禁止主义，无论是自然还是人工的色彩，历史告诉我们，对于设计师而言，物品的色彩是一个重要的主题。今天，化学颜料使得色彩更加缤纷，电子技术让色彩复制再生，这个观点也日趋明显。

教学方式 > 包豪斯帮助设计完成工业产品向美观与功能的转型，同时也提高了大众的审美品位。包豪斯以实验为基础培养预备设计师，他们通常以技术与艺术为背景，同时具备团队合作能力，团队合作让他们能够掌握必要的多样性技能，保证在生产时，为功能复杂的技术问题提供解决方案。但也需要同时掌握平面艺术、设计与建筑的横向关系，来共同定义产品的形式。在包豪斯，许多现代建筑大师进行设计教学，这绝不是偶然。许多建筑与设计之间的风格如此相似，这也不是偶然。在欧洲，至今为止，许多设计学院依旧与建筑专业或视觉艺术专业相联，也绝非偶然。它们深受包豪斯影响，因此更加倾向于设立与组织工业生产相关的课程，其中这些涉及到技术创新与产品性能。马歇·布劳斯（Marcel Breuer），艾琳·格瑞（Eilleen Gray），赫雷·厄斯特（René Hernst），密斯·凡·德·罗（Mies Van Der Rohe），让·普鲁韦（Jean Prouve），柯布西耶（Le Corbusier），瓦尔特·格罗皮乌斯（Walter Gropius）这些大师在设计领域，奠定了法国及中欧地区设计学校的基本特征。无论欧洲还是全世界，只要是他们工作过的地方，都留下了深远的影响，这更加证明了现代运动的意识形态形成了设计的社会职责的基础。

50年代后，乌尔姆（Ulm）学院进一步丰富了设计内容，完善了设计课程。它开设了视觉传达和电子信息的理论课程，以及对产品操作与管理的实践课程，来完成产品的快速实现，以防过时。

可持续技术 > 多年来，设计实验越来越普及，设计师和企业也开始提出产品的可持续发展。这类产品适度使用自然资源，合理利用材料和生产流程，借助其形式与功能来优化产品的操作时间。毫无疑问，实验领域充满了创意，与过去相比有很大进步，也更加注重方案的形式内容。但是它们依旧无法摆脱过去的影响，无法摆脱那些与城市和家庭空间概念相关的类型范式和生活方式，即使那些已经老旧过时。事实上，依旧有不少设计师认为可持续发展技术会妨碍影响建筑，他们担心新的方案会颠覆传统的空间概念及结构形式。

提到环境的可持续发展，我们必须直面新技术，无论是有形的还是无形的。这些新的技术会带来真正的设计革命，促使产生新的居住模式。总而言之，对涉及可持续发展的主题，建筑师与设计师可以做的有很多。一方面,他们可以积极推动住房概念,让房屋成为不仅"消耗"同时也"生产"能量的机器。另一方面,尽管在产品与建筑概念上,模式与美学方面的范式焕然一新,他们似乎也无法形成更新更大胆的尝试。

可持续发展技术这一方法普遍存在,它或多或少涉及所有设计领域,当然也包括工业产品设计。比如,当智能显示器感知用户远离时会自动降低亮度;吸尘器吸走灰尘时会自动压缩减少使用空间;手机由可回收材料生产会减少能源的浪费;冰箱使用特殊的独立面板控制使得绝缘效果高达 20 倍;汽车的排放量将逐渐减少到零。然而,对于产品的形式与使用方式,依旧无法完全正确地形成全新的可持续发展的世界,而这正是可持续发展产品系统应当面对的。这样看来,即使我们探讨在数字技术的支持下,可以由节约能源或实现能源转化的需求引起创新技术因素,我们也并没有完全意识到,我们正走向一个新的概念世界,迎来新的时间与空间的关系。当初在工业革命结束之时,包豪斯为了新的美学革命,试图赋予时代技术形式一些特殊意义,现如今,运用新型技术的设计师也正处于同样的境况。

连接 > 当下,我们可以设计环境。所谓环境,是指人类完全沉浸其中的、未来的可触摸的无线电空间。无线电空间的原型 Tune 是由伊夫雷亚(Ivrea)交互设计研究所的设计团队研究完成。它是可触摸的,人们借助它探索新的情感、视觉体验,感受新的空间的变化。当选择不同频道时会响起音乐,伴随着光线会显示出各种各样的图像。同时,在阿尔斯特大学开发完成的 IMP(Individual Memory Projector)原型,也就是我们的"记忆盒子",它形似球体,中间有个平面与其相交。将它旋转打开则可变身为一台投影机与人类进行互动,它将人类的记忆通过视频及图像的形式,投影至任一白色平面上,并伴随着声音和气味,唤醒我们的记忆。

匹兹堡的卡内基梅隆大学的研究学者们实现了 The Hug 原型。The Hug 被称为爱的枕头,实际上,它是拥有柔软臂膀和脑袋的机器人。当你通过它进行通话时,它会让你感觉与对方十分亲近。在通话过程中,它配备的软件可以识别声音或者其它信号。通过光线和热量(由特殊热能纤维产生)或着通过震动将双方设备接收到的感官数据相结合,来实现当你对着麦克风说话时的人机交互。

通过电发光,Oled(Organic Light-Emitting Diode)技术研发完成了超薄、超轻、可弯曲的屏幕,且色彩丰富,具有出色的亮度和清晰度。这些屏幕可应用于电脑、电视及手机上,在使用过程中还可以实现卷屏。

交互设计是一个应用领域,在该领域中,无处不在的技术被注入到我们周边的一系列人工产品中。这些技术虽然不显眼,却在我们的日常生活中触手可及,无时无刻地影响着我们的行为,刺激着我们的感官,给我们带来了新的愿景,为我们开启了全新的体验。这些技术促使我们与人工世界进行交互,在改变产品的同时,也改变了我们的情感或需求。交互设计,即通过软件技术的使用来设计简化人与物之间的关系。因此,这需要设计者掌握一系列复杂的能力,其中包括对软件与硬件的技术工程、通讯语言、社会学、人机工程学以及心理学的了解。交互设计领域中诞生的产品需注重新型技术，直面无形的虚拟世界，具备全球通用的图像语言。交互产品设计在使用日益发展的最前沿技术的同时，也让我们反思由此产生的强烈的依赖感。当我们谈及的产品大多为无形时，只有人类与其进行交互，才能赋予其生命和存在的价值。在某些情况下，通过人类的干预，产品唯一的存在感则是展示了自身的功能与用途。人类介入的动作甚至可以直接决定它们的造型、形态、声音及气味等。但事实上，产品中既定了软件编程以及配置管理，此时人类显然只是起到了表面的"主动作用"。

含糊的概念 > 随着时间的推移，各种新型材料也得以展示其真正的属性。如今，不再是木材的语言更加精细而塑料的语言相对粗俗。木材的语言简单明了，描述了人类通过器具对木材进行切割、刨平、打磨及抛光的过程。塑料的语言暧昧隐晦，体现了人类通过智慧将化学物质转化为材料的这一封闭含糊的过程。塑料即使在生产完成之后，依旧能够与人类、时间及环境相互产生作用。多年来，一方面存在这种"表面的怪异"，另一方面也具有这种"含糊的概念"。也就是说，设计师使用新型材料来惊艳亮相，运用对它们外表一致性的理解使人产生错觉，试图建立人造与天然制品之间、日用品、技术产品以及智能产品之间的新的关系。慢慢地，他们开始使用新型材料的多面性来对抗传统材料简单明了的语言。这些新型材料总是更加轻盈、易变、透明、高科技、光亮、反射、冷峻，像变色龙一般，具有欺骗性。如今，当我们再次提及自然，不再是简单地模仿外表，而是更关注其内在特性。如何通过材料和产品来改变其感官印记，甚至涉及生态仿生及生态模拟。所以，我们应当更加注重材料的自然特征，而不是简单地关注其外表。

教育经历 > 在复杂丰富的生产系统中，涌现了众多年轻的设计力量。即使这些年轻的设计师们在有些情况下依旧停留在传统的设计领域，设计那些并不是以高性能和高技术为追求的产品。在他们之间，矫饰主义语言的实验风格尤为盛行。这促使学校越来越重视新的科学技术领域，将新技术转化为新产品，从而提高人们的生活质量。同样的，年轻设计师更加注重在复杂创新体系中的所有"成员"。在此，抛开过度设计的时尚主题，设计师们应当具备分析设计周边现象的能力，而时尚主题也只能帮助设计扩展传统以外的概念。对于年轻设计师而言，学校需要提供更多的方法来完善他们的教育经历，使他们更具综合性。以便帮助他们在工业设计领域中，能够满足各种高水平创新技术的设计需求。

市场需求 > 新型技术参与生物加工，创造转基因生物，从而将人造及工业产品的概念延伸到食品行业。比如西红柿、西葫芦等，过去只能任由其自然生长，而如今可以在生长周期中改变它们的颜色和大小来满足不同的市场需求。

新型技术让产品更加智能,它们可以与人类互动交流、传递情感、实现远距离倾听、增加幸福感,使人与人之间的交流更加便捷。新型技术不仅可以将产品用于身体外部,还能将新的功能直接植入我们的身体内部。比如让我们得以在黑暗中观察,在水中诉说与聆听,让我们的视觉超越了真实的维度。

惊艳亮相 > 如今,对设计师而言,不再是要设计满足特定需求的产品,而是要设计能够让消费者产生购买欲望的产品。设计师描绘的产品与实际需求的产品完全不同,极具象征意义。或者说他们促进了产品的非物质化,使得产品逐渐成为一种沟通方式,一种具有自身标准及明确语法的语言,这些产品不再是以满足某种需求为最终目的。从一方面来说,这种现象让"创新思维"不受任何物质限制,成为一种自由的纯粹的思维概念。另一方面,设计师的社会作用,被产品的光芒所掩盖,似乎无法避免地消融于一个光亮的世界。如同所有的因果关系,从产品的惊艳亮相到设计师的惊为天人,这一过程极其短暂,设计师也逐渐闪亮登场。作为明星体系的真正的主角,设计师应当具备管理自身媒体价值的能力,善于推广自己的个人形象。因此,设计师逐渐成为人们追逐的明星,成为一个可以被触摸、被观赏乃至被聆听的具体人物。明星之中的明星,比如飞利浦·斯达克(Phillipe Starck),他本人就是一个品牌。

沟通交流 > 一个设计师应能够轻松游走于现有的媒介工具中。从纸张到屏幕,从机械到电子,沿着传统,我们以古老的图文形式为基础,感受从15世纪中期诞生的印刷艺术,直至21世纪互联网技术的来临。我们不得不重新思考,近年来平面多媒体设计师在其领域进行的操作、实验和研究,开始涉足工业产品设计领域,这两大学科开始构建共同的设计学科框架。这种情况,也可以引用教育部颁发的设计学科介绍来进行阐述:"设计学科涉及的理论与方法;工业产品(现实与虚拟)设计的技术与工具;在产品生产过程中的技术构成特征;产品功能性;在空间环境关系及工业市场关系中的使用形式及功能。这些产品的本质(从消费品、生产资源到耐用品,从通信交互用品到关系与服务系统)及其复杂性(从材料半成品到中间产品,从零部件到最终产品,直至与产品、通信、服务相关的交互系统)表明,作为跨学科专业,设计学科涉及许多生产方法与技术,与不同的市场生产部门相互作用,使得特定研究领域得以不断发展。"事实上,这两大设计学科涵盖了不同领域的知识与技能,独特的设计理论知识,以及以试验与实践为基础的技术工具。一方面,平面多媒体的相关理论与实践正逐渐融入产品设计,与其设计文化相结合并形成共同点。另一方面,这两大领域各自的独特性也是无可争议的。正因如此,我相信,把这两种学科视为工业设计领域的一部分是合适的,甚至是必要的。同时,必须强调这两个领域的实验与科学研究应该保持在特定的范围内,它们相近且有共同部分,但是却各自拥有独立的学科界限。

错觉感 > 即使是大规模生产的工业产品,也应当思考它的短期批量生产,即将产品分阶段生产。在

构思消费品

保持功能部件不变的同时，利用分段生产，不断地改变它们的形式及外观。这样的思维也转移至建筑领域，事实上，建筑制造业正迅速地从工业制品生产发展的大量研究中获益。建筑作品运用产品的实验技术，变身为一台具有美学价值和工业产品复杂性的"机器"。比如，建筑物在逐渐非物质化的同时，由丰富的电子机械部件构成并改善其性能。如同工业产品一般，建筑制造所展示的通信符号，使其更具附加价值，但同时也缩短了使用时间，进而加快了建筑物的老化过时。此外，被广泛使用及改良完善的结构框架，有效地将建筑外壳与支撑结构分开，并在构成骨架外形成一层表皮。除了技术上的转移，还有执行方式、性能、象征符号，如今，所谓的"大都市怪物"也就是将工业产品的所有需求转移至更大比例的建筑物上。过去，场所一直是建筑作品中至关重要的一环，但是如今也慢慢失去与其之间的关系。最终，所使用的形式与材料都没有与物品的目标用途形成一个有机关系。在产品及建筑设计当中，模糊性与错觉感的比重逐渐上升。它们的最终使用用途被隐藏起来，而广告形式更加突出了这种模糊性。产品因为自身的信息内容变得更加复杂，与建筑物一起，使得城市本身呈现出一种模糊的形式，并成为沟通交流与公众信息的临时载体。矛盾的是，这些工业产品与建筑物所使用的技术越是先进，它们存在的时间却越是短暂。

有关材料的思考 > 当我们谈及周围的事物时，它们与材料之间的关系让我们的记忆不断涌现，并随之产生了一种强烈的情感。这么说来，材料是所有产品的"构成文字"，材料让产品更加具象。同时，人们过去及现在依旧使用各种材料来制造不同的产品。随着时间的改变，材料由之前的手工加工，发展到由机器进行塑造和修改来不断地适应生产需求，但同时，材料也塑造并改变了手工技艺和生产技术。材料存在于艺术之中，也是艺术的前提条件。因此，它能够直接唤醒材料所涉及的文化。比如，石头、粘土、木材之所以被选为建筑房屋与构造物品的材料，不仅因为它们是可用的自然资源，还因为它们的建造功能，它们具备的静力学、持久性、可操作性和与其它材料结合的能力，以及它们的可塑性和美观性。让·努维尔（Jeann Nouvel）为巴黎拉德芳斯区设计的《无尽之塔》，是一个极具象征性的作品，它表现了超高智慧，体现了材料模糊性与简单性的双重语言。这座塔呈现为完美的圆柱形，从地下深处延伸而上，天然的黑色花岗岩岩石，呈现出一种"真实性"，直至消融于地面的玻璃中。地面十层以上，花岗岩显得十分光滑，在它的表面折射出天空的光芒与色彩，使其具有一种偶然的"模糊性"。当它进一步向上融入黑色玻璃时，错觉感就逐步形成了，花岗岩和玻璃合为一体并逐渐非物质化。位于更高处的玻璃出现印花及哑光，让一切短暂易逝，出现了不确定性。最后，在云层中，玻璃再现其透明材料本身最真实的特征，与天空融为一体。这一设计作品表明，新型材料可以在设计语言上开辟一个非凡的实验领域。

磁卡 > 设计，可以让有形产品形成无形关系。比如，一张小小的智能卡片，以极小的体积装载了丰

§ 设计_文本与语境

富的内容,通过计算机算法,以无数的链接将我们置身于一种无形的关系网络。日本发明家有村国孝(Kunitaka Arimura)将自己的想法应用于塑料卡片,并在其中植入了能处理并储存数据的微处理器,智能卡应运而生。这些智能卡片好比真正的电子保险箱,它们不仅可以作为信用卡使用,还可以被当作电话卡、安全门的钥匙、个人信息记录卡以及数字电视访问卡使用。智能卡是改变我们行为习惯的最重要的因素之一。

磁卡的呈现多种多样,如电话卡、徽章、信用卡等。因此,思考磁卡在我们的日常生活中所具备的"物品价值"显得十分重要。事实上,它之所以是一张小小的卡片,这并非偶然,它的尺寸与其说是我们决定的,不如说是与其相互作用的机器决定的。但无法避免的是,磁卡的出现,也一再改变与之配套的物件的尺寸,如钱包、卡套、口袋等等。另外,智能卡代表了典型的新的产品设计。像是由磁卡展开的供应商与服务用户之间的交互设计。或是超市和加油站提供的消费积分的累积奖励,又或者那些奢侈品牌(Prada)的客户身份识别卡,根据客户登记的个人信息,尽可能地利用定制来满足他们的需求。小到个别产品概要,大至企业战略布局,智能卡都能找到用武之地。新的产品设计更加注重材料与新型科技的结合。与纯粹的形象相比,会更加重视响应感官的需求,如触感、愉悦度和舒适度。例如,在Technogel 凝胶枕案例中,飞利浦·斯达克(Philippe Starck)和盖特诺·佩斯(Gaetano Pesce),将生物医学材料应用于产品中,使其具有柔软性及保护性。这类产品通常能给人以愉悦感,满足人们的情感需求。这也满足正如营销专家费斯·帕帕考恩(Faith Popcom)在市场趋势中所定义的舒适感。

同时,设计利用新的方式,新的信息工具手段,与生产过程紧密相联,从而在设计基础与实践中产生真正的变化。

有益的陈旧 > 另一种管理人工产品生命周期的方法是延缓其衰老。即替换或加强产品的技术部件使其性能与技术发展得以兼容;又或者通过同型装配,选择仍然有用的部件来实现新的组装产品;在特殊情况下,甚至通过去除部分零件来获得一个残缺但仍然可用的物品。

但是依旧会出现产品自身的功能逐渐减弱,以至于被其它功能占据使用的情况。有多少次,我们在花园里看见食品罐头和破旧的鞋子变为花盆?又有多少次,我们看见废旧报纸变为食品包装袋或泥瓦匠的帽子?它们仿佛一面现代化的镜子,映射了"资源匮乏文明"的创造力,人们会根据自身的经验物尽其用。就像马路上的孩子们,他们会利用废弃的材料自制玩具。他们用旧木板制作滑板车,用罐头瓶和铁丝制作车厢,用自行车轮胎制作铁环,用塑料袋制作风筝等。

面对功能衰退的产品,我们可以用一个新的层面来进行表达。我们将"无用"的东西置身于新的场景中,甚至会通过新的视角呈现出意想不到的美丽,且更具诗意。

比如,被遗弃的汽车影院的屏幕摇身一变,成为了与风景相映成趣的巨幅照片,好似爱德华·霍普(Edwad Hopper)的画作一般。古老的加油站,孤立的仓库,被遗弃在农村的土地上,毫无用处。粮仓与油槽车，似乎是广袤空间中雄伟的雕刻艺术品，唤起了过去的功能与传统，它们可被升华为具有意境的遗址。遗弃的老旧货车表面，厚重的钢板被时光侵蚀，锈迹斑斑，它坚固的结构提醒着我们它所经历的漫长旅行。铁路与货运公司的招贴海报如今也只剩下浅浅的色彩痕迹。曾经的人造物仿佛雕塑一般，成为自然界的一部分，让我们深深地感受到它们的魅力及其灵魂的力量。

俏皮亮相 > ToolToy，是一种日常用品设计生产的趋势，即对待产品如同对待玩具一般。讽刺的是，在这些产品极具表演性质的外观形式和结构美学的背后，隐藏着真正的功能用途。它是一种风格趋势，来源于对产品沟通内容的附加价值的认可。事实上，它不再是由美学与性能之间的平衡来决定，而是通过事物形式的绝对价值进行表达。近年来，通过对高度创新技术及新功能的使用，工业产品逐步实现了非物质化，并成为深入美学研究的"累积体系"，这一体系既不实用，也不能满足最基本的需求。随着先进的机械技术在日常产品中的应用，我们实现了产品的高效性和关联性，而卓越的性能也日益成为消费者选购产品的标准之一。随着印刷电路的普及，电子信息的应用，卓越的精密性能已经变得无法超越。因此，除去性能因素，产品的附加价值逐渐由其自身的形象价值、设计感及象征意义来决定。

事物、物品与产品

复杂流程
摘自《设计无处不在》，2013年1月讲座

从未间断
摘自《工业的自然属性》，2012年《DIID》第55期

激发创造力
摘自《玩具好似欲望物品》，2012年《DIID》第54期

建筑物品
摘自《规模与当代景观》，2007年《DIID》第31期

战略价值
摘自《健康设计》，2007年《DIID》第27期

为运动设计
摘自《设计表现》，2006年《DIID》第20期

无名但工业化
摘自《众多或无名产品的力量》，2005年《DIID》第15期

物品与环境之间
摘自《污染、转移及认可》，2005年《DIID》第14期

真实性与模糊性
摘自《材料的真实性与模糊性》，2005年《DIID》第13期

技术创新
摘自《高科技》，2004年《DIID》第9期

杂交产品
摘自《伦理 & 技术》，2003年《DIID》第6期

设计博物馆
摘自《感觉的转移》，2003年《DIID》第5期

升级
摘自《全球设计》，2003年《DIID》第3—4期

当我们阐述人工世界时，我们常常用绝对简单的词汇来描绘无处不在的事物、物品与产品。这些词汇往往有着相同的含义，同义词一般地描述我们所处的环境。但是它们并不完全重叠，它们既不是文学语言也不是设计语言。尽管如此，无论是日常活动中的随机性还是它们自发的确定性，是我们斌予了这些简单词汇以描述产品的尺度及作者的含义。这种阐述方式确定了工业设计的本质，也就是知晓物品涵盖一切范畴之后将事物转化为产品的过程。

复杂流程 > 在手工艺生产过程中，即使属于相同的种类，具备同样的形态特征，也能诞生出独一无二的产品。因此，同系列产品的不同之处，可能是因为一系列的控制的不同，如材料按压力度的不同，雕刻工具力道的不同，颜色组合选择的不同，手工匠人经验与能力的不同，甚至是他们当下感受的不同，因为他们对作品形式或功能与当下情感的联系有着绝对自由的创作空间。相反，由先进的工业生产技术制造而成的产品则不存在这样的缺陷，其功能操作部件是由电子电路构成，借助微晶片赋予产品本身以生命力。这意味着我们使用的产品可能诞生于这两种不同的生产模式。一方面，我们不能忽略个人在特定条件下的创造力，其技术能力、手工能力、工具使用的熟练度、传统或创新能力；另一方面，在流程复杂的生产阶段，只有部分程序由人控制，人们借助机器的强大技术，来呈现完美且高性能的产品。后者当中，这些人工造物由高精机械分为几个阶段生产，通过远程操作机器进行生产来实现生产制造阶段与前期设计阶段区域上的分离。同样的逻辑，质检部门制定计划与实施的地点也可以有所不同。重要的是，正是生产过程中通过对器械的使用，才使产品的大量生产与个人定制得以并存。

从未间断 > 几年前，一篇题为《伟大的设计图集》的文章开篇便写道："从清晨的第一个姿势开始，我们伸出手去关掉闹铃，直到晚上，我们关掉床边的台灯（或关掉电视屏幕），我们一直沉浸在产品的世界当中。这些产品支撑着我们的生活，迎接我们，装点我们，帮助我们却也妨碍我们。如今，这些丰富的产品已然以一种想当然的方式充斥着我们的生活，看似十分自然。尽管我们对产品产生兴趣，并花大量的时间去选择或整理它们，我们却很少仔细思考这些产品的本质，以及在不久之前产品稀缺、珍贵、难以获得和保存这一事实。在很长一段时间里，只有国王、牧师、将军和富商这类少数权贵可以使用这些物质资源。而其他人只配有办公所需的工具和少量的必需品。"这段摘录的文字，阐述了我们熟知的设计现象，是对完全被设计史上不断产生的物体填满居住空间的真实写照。面对这种不可阻挡的由消费带动的产品流，我们偶尔会习惯性地近距离观察这些生来不会过时的产品。这些产品并不局限于说明书，它们的名字往往遵循其功能，及其明确的构成材料。而今，我们正处于无名设计领域。我们并不想这么称呼它，但是我们依旧受到批判叙述模式的制约，并没有找到完全有效的替代说辞。事实上，威达（Vitra）公司尝试了以《设计隐藏的英雄，日常事物的天赋》为题的展览，它在伦敦科学博物馆展示持续至2012年6月。从展示的这一系列的物品中可以明显看出，人类总是能够通过创造力，谨慎地将纯粹的智慧与设计结合在一起来满足自身需求，而英雄主义则一直存在于自主创意及社会创造力当中。

激发创造力 > 与其他物质文化现象一样，随着时间的推移，社会和经济背景逐渐发展，玩具以一种真实且极具表演性质的形象出现。众所周知，儿童通过玩具可以模仿成年人的日常行为。他们小范围地将家庭空间内外的物品进行模拟、缩小比例,来模拟对应的成人的行为模式。

玩具一直扮演着教学角色。从麦卡诺(Meccano儿童钢件结构玩具)到乐高,再到摩比世界或者那些拼图游戏,搭建玩耍结构玩具能够不断地激发孩子的设计创新力。孩子们的创造力还可以通过各种不同的玩具得以发展,其中包括具有怪异人物形象的故事、漫画和卡通。玩具中的设备能保持孩子们的活跃度,比如手动充电或电池供电的机械物品,以及可以进行人机交互的电子信息产品。我们回顾玩具历史的演变,它所涉及的人类学、文化及社会体系,影响了产品的成型模式,也改变了它们的制作工艺,它们逐渐由传统手工转向工业生产,由人造材料替代天然材料。围绕着玩具的生产,许多公司应运而生。在此,真正有品牌的产品则具有和成年人产品相同的特征,也就意味着产品都是从设计师概念出发,由典型的工业程序进行生产。正因如此,这些产品也实行时尚营销,将产品销售至世界各地。

比如,费雪(Fischer Price)公司成立于七八十年代,在设计师爱德华·萨维奇(Edward Savage)的主持下确定了自己的国际地位;沃尔特·迪斯尼(Walt Disney Pictures)公司(如今的皮克斯动画工作室Pixar Animation Studios),于三十年代末期已经制造了第一批动画产品;美国女孩(American Girls)则成为受到当地女孩与青少年极度追捧的著名玩偶制造商。

设计在玩具领域扮演着越来越重要的角色。这一点从2001年Bozart Toys生产制造的万花筒之屋就可以看出。它由劳瑞·西蒙斯(Laurie Simmons)和皮特·维尔莱特(Peter Wheelwright)提出概念,是一所墙面透明且色彩斑斓的住宅,旨在为孩子们提供一种新的游戏方式,最重要的是将设计与艺术品对比的概念从小融入他们的生活。万花筒之屋的配件套装也都是出自大家之手,贾斯珀·莫里森(Jasper Morrison)和罗恩·阿德拉(Ron Arad)设计了客厅的装饰品;凯瑞姆·瑞席(Karim Rashid)设计了餐桌;劳瑞·西蒙斯(Laurie Simmons)设计了卧室的床铺。而房间里的艺术品则由巴巴拉·克鲁格(Barbara Kruger)、辛迪·雪曼(Cindy Scherman)、彼德·哈雷(Peter Halley),以及艾伦·麦克勒姆(Allan McCollum)等艺术家完成。当然,这所房子由穿着时尚的蓝绿家族长期居住。

建筑物品 > 在产品、建筑、城市环境相互之间的关系中,由于逻辑从起初的"技术转移"到后期的"风格同化",工业产品的"巨人主义"得以加强,但这实际上是一个反渗透。说的是建筑物品运用产品的实验技术,变身为具有复杂性的工业产品。它在非物质化的同时,如同现代工业制造的产品一般,逐渐建立其通用符号价值。由建筑制造展示的"消耗时间"及"计划废弃",然而,原先都只属于最先进的商品工业理论。

战略价值 > 随着全球市场竞争的加剧,发展中国家需求的增长,为了提高竞争力,通常需要增加产品的绝对数量、类型及性能,提高产品的外观及标志性水平,而后者往往在竞争激烈的市场中起决定性作用。在此,企业发掘设计优势,首先将其视为创新商品的战略价值,其次为产品体系及生产过程

事物、物品与产品

的发展及把控起到了决定性作用。这一设计结合形式特别适用于意大利的复杂生产系统，它逐渐成为工业制造区域变化中的良性模式。

为运动设计 > 所谓运动设计，就是将运动领域的需求阐述且转化为产品，并不断地寻求其性能的极限。因此，这需要先进且持续不断的实验。这些产品必须有助于提高运动员在其领域内的表现，同时也必须保护他们免受各种风险伤害。举个例子，为了帮助足球运动员托蒂（Francesco Totti）伤后重返世界杯，迪亚多纳（Diadora）研究中心专门为他设计并打造了一个护胫，用来保护他四个月前受伤的左腿。这是一个极具解剖学和人体工程学的产品：三层碳纤维保护腓骨区域，外部由钛制成，内部则为纤维B构成，且直接接触运动员腿部。这只是其中一个大家熟知的例子，借此表达在运动设计领域中同样存在着工业设计的丰富性与复杂性，说明如今可以通过特别的设计或色彩，又或通过建立特定的解剖学特征，来定制产品满足用户的特定需求。

无名但工业化 > 有些产品与设计文化紧密相联，同时，也存在另一些大规模生产但拥有普通本质的工业产品。有些"著名"产品由大师设计完成，因此被视为实验文化的象征，或因其特定的时代特征而成为备受追求的物品。同时，也存在无穷无尽的用于大规模消费的"无名"工业产品。然而相对于那些"文明"产品，这些繁多的日常用品，反而更能见证工业生产的必要性。它们代表了工业社会、后工业社会，以及今日的全球化社会的物质文化。它们是无类别产品，也与使用语境无关，相反，慢慢地，它们更加倾向于代表不同的经济阶段及个人和社会的特定行为，成为对大众审美的示范展示。这些无名产品帮助我们重新思考了与工业生产内部矛盾相关的主题，让我们了解了设计和量产之间的关系。伴随着经济、政治和文化模式的发展，这种关系迅速生成，同时也与所处社会产生紧密联系。正如福特主义、理性主义及功能主义时期的模式，它们所处的社会依靠普及的消费逻辑，促进形成民主意识；消费品的计划废弃模式，看似增大了社会的财富价值，但实际上持续不断的需求刺激了更大的消费；波普文化时期，通过对大量的生产模式的抨击，试图重新找回个人自主的价值；时至今日，由于电子通讯的飞速发展，全球化的来临，调节产品流的时空不复存在，因此，其中的文化、传统特征，使用方式以及社会关系也都逐渐消失。

物品与环境之间 > 风格研究和技术创新贯穿于日常用品和其它艺术表现形式当中。正如索耐特（Thonet）椅的诞生得益于木材弯曲技术的实现。19世纪末期，建材技术迅速发展，建筑领域深度革新。其中，水晶宫就是一个极具代表性的例子，它通过使用预制方法，使得合金及玻璃材料的使用成为可能。麦金托什（Mackintosh）设计的椅子，亨利·凡·德·威尔德（Henry van de Velde）设计的小物件（陶瓷、餐具等），维克多·霍塔（Victor Horta）设计的手柄、柱子等家用或城市空间物品，它们都借助蜿蜓的曲线阐述了与大自然的关系，与新艺术运动息息相关。所有风格主义所处的时期是一段多元化的历史阶段，它将家用办公物品以及建筑城市环境相结合。正因如此，我们可以通过里特·维尔德（Gerrit Rietveld）的红蓝椅，借助位于乌特勒支的施罗德（Schroeder）的住宅，以及彼埃·蒙德里安（Piet Mondrian）的画作重新解读风格派。马歇·布劳斯（Marcel Breuer）的第一个系列的木制椅受新造型主义影响。1925年马歇·布劳斯（Marcel Breuer）与1926至1929年间密斯·凡·德·罗（Mies van der Rohe）都设计完成了金属管状椅，他们利用了新型金属合金的特性，同时也为密斯·凡·德·罗（Mies van der Rohe）的巴塞罗那馆确定形象布局及几何空间形成了有用的参考。柯布西耶（Le Corbusier）的可调躺椅象征着大师们对灵活性进行的长期研究。阿尔瓦·阿尔托（Alvar Aalto）利用胶合板技术，并在心理学领域研究对应的舒适度和反应性；或是小沙里宁（Eero Saarinen）对有关木材的化学处理的研究，或是使用大型压力机来制造扶手椅的"有机"造型，他们所做的这些都与有机运动的核心基础有关。阿诺·雅各布森（Arne Jacobsen）的模块化椅子，利用木材塑化，发展新的表现主义形式。随着时间的推移，物品与其所处环境之间的关系逐渐发生概念性的改变。起初，这是一种技术转移的关系；之后，慢慢变为一种风格同化的关系。

真实性与模糊性 > 如今，我们可以通过对称的表现形式来描述当代材料的使用。除了传统的用途，还存在一种新的方式，这是一种值得研究的现象。可将其概括为：材料的"真实性"与"模糊性"。显然，这种新的方式得益于材料研究产生的不断创新。现今，我们探讨超性能材料，它们更加轻便、耐用和灵活；我们探讨多维材料，它们是在物理空间里定义的三维材料，能够生成具有未知特性及外观的物质；我们探讨再利用或重组的材料，回收、混合、转化的材料，在不同的场景下，用不同的方式，它们几乎可以替代原材料，或者更简单地重新组合；我们探讨智能材料，它们能够与外界因素相互作用，与我们的感官进行互动并在预测控制下产生改变，又或引入类似于数字化信息的材料科学发展成果，使得他们能够在纳米水平上提取生物系统DNA并对其进行设计改变。

因此，材料具有的附加价值，不仅存在于表现形式当中，同时还体现在与感官相互作用的能力上。材料产生新的语言，能影响人类的行为，改变物品的使用方式，以及与复杂人类的感官交互的方式。

技术创新 > 工艺与人工造物的新的概念开始发挥作用。过去，"工艺"是指使用合适的工具与程序来简化与完善产品的生产，"人工造物"是指能够直接满足人们需求的产品，无论功能如何，都为了改善人们的生活条件。而将需求转化为产品的过程是人类创造力与技巧结晶的过程。借助创造力，人们设计满足需求的产品，并定义产品的性能标准。利用工艺与技巧，人们按照预先设定的目标筛选适于产品生产的材料、工具和操作方法。在工艺与人工造物的关系中则体现了发展本身。事实上，创新技术促进了科技产品生机勃勃的发展，迄今为止，这些科技产品作为身体的延伸物，以一种更精确更有

事物、物品与产品

效的方式执行具体功能，其中包括人类的感官功能。如今，除了在家庭环境及城市空间中可见的科技产品以外，新型技术还涉及无形产品。比如，通过生物学变化来改变我们的身体，借助产品来丰富人体的功能。新型技术还参与生物加工，创造转基因生物，从而将产品的概念延伸到食品行业。比如西红柿、西葫芦等，过去只能任由其自然生长，而如今可以在生长周期中改变它们的颜色和大小来满足不同的市场需求。新型技术让产品更加智能，它们可以与人类互动交流，传递情感，实现远距离倾听，增加幸福感，使人与人之间的交流更加便捷。新型技术不仅可以将产品用于身体外部，还能将新的功能直接植入我们的身体内部。例如让我们得以在黑暗中观察，在水中诉说与聆听，让我们的视觉超越了真实的维度。重新思考人类自身的状态以及和未知难测的科学的关系，显得尤为重要。技术创新具有的强大潜力，正推动着我们超越所有可能。

杂交产品 > 我们正面临着显而易见的两大方向。一方面，有些产品最大的吸引力是其唯一性，这一独特特征取决于它们的手工生产过程。这些产品往往被视为"民族产品"，因为它们形成了一种物质文化，也展现了一个民族的文化和历史。另一方面，有些产品被大规模地加工生产，它们试图在性能与形式上追求一种完美，理论上来说这些产品独有的特征是对创新技术的使用，也因此可以称它们为"技术产品"。除此之外，出现了一些可以替代民族或技术概念的产品，它们外在形式类似民族产品，但却借助新型技术实现其生产过程。混合文化的现象由此诞生，它产生并推动产品的"杂交"属性，使得它们既不是"民族"的，也不是"技术"的。当真正意义上完全由传统手工制成时，它们是"百分之百民族"的；当使用生产技术产品的机器来仿制传统手工时，此时得到的产品是"民族"与"技术"杂交的；当产品充分发挥先进技术的生产工艺时，它们是"百分之百技术"的。这种发展状态具有重要意义，同时也说明了混合现象已然成为全球风格的主要特征。

设计博物馆 > 如今，越来越多的技术产品变身为教父产品而被收藏于设计博物馆中，这值得反思。借助机械、机电、电子部件赋予技术产品以生命，通过它们发出的声音、图像及运作，这些产品与人类进行交互。因此，这些部件是技术产品的本质，当它们受损时，产品就没有生命。这样一来，设计博物馆还有意义吗？收藏并展示失去原有功能的大量生产的技术产品，将它们奉为经典是否有意义呢？在设计博物馆里，我们可能会在橱窗里看见安静沉默的老式收音机排成一排，镜头封闭的照相机整齐陈列，毫无生机的打字机规整划一。展示物品的使用年限越是临近，越会给我们一种错觉，仿佛我们并不是在博物馆里，而是置身于旧货商店中。

事实上，我认为应当由科技馆记录并展示工业产品及其体系的发展。与其在一堆沙发、书桌、电视机中展示一件死气沉沉的Lettera 22打字机，不如在某个地方静静地欣赏它们的技术性能的发展变迁。1753年，在维也纳的实验室中，弗里德瑞秋·冯·莫林（Friedrich von Knauss）第一次想到用机器进行书写；

慢慢地，打字机进入工业化生产阶段（1873 年的 Remington）；紧接着，第一台字母可见的打字机（1896 年的 la Underwood n.1）诞生；第一台电子打字机（1901 的 Cahill Writing Machine Co.）诞生；直到发展为目前的智能机器。在我看来，如果技术产品具有形式美学价值，那么它们可以以相关主题形式或者以对特定风格及设计语言的探索在当代艺术博物馆进行展出。我想，如果这个想法可行的话，我们应该进行深入的探讨。

升级 > 随着全球化的发展，产品和货币已经在全球范围内循环流通，各大品牌开始经营企业的社会形象使其统一化，让大众逐渐成为其消费者。借助某些产品可以很清晰地阐释这种现象，比如可口可乐，牛仔裤，以及从任天堂的GC到索尼的PS的最新一代游戏机。

可口可乐公司借助其有形的产品，无形的品牌及其传递的信息，红底白字的品牌符号，渗透进两百多个国家和地区，几乎拥有全球的市场。正如让·鲍德里亚（Jean Baudrillard）所言，它是零度软饮料的象征。统一的语言，蔓延的广告，难以用其它语言表达的"可口可乐"，让它成为一件教父级产品。它从诞生伊始就是一个极具沟通意识的标志，当时"可口可乐"这个名字被认为是最基本的，因为它在几乎全球的语言中有着相似的发音。从来没有一个产品有如此多的广告，从 1886 年可口可乐在亚特兰大的约翰·彭伯顿（John Stith Pemberton）药房里首次面世，直至 1908 年，22 年的时间里，可口可乐占据了 250 万平方米的广告空间，它的商标在过万的商店橱窗里出现过。

牛仔裤是跨越性别和社会地位界限的标志，这种服饰不再属于某种特定的文化，而是代表所有。最初，牛仔裤除了反对分化以外，不具备任何意识形态的信息。但之后，它却违背了自身的原则，通过不同的剪裁、蓝色调、拼接和搭扣，它逐渐成为一种时尚的产物。但是这种对新式的追求并没有产生什么实质的作用，牛仔裤的版型、质地、饰钉以及织品，只是保障了它的老化更优，以及通过洗涤对色彩进行的不断改善。而索尼最新一代的 PS 游戏机展示了全世界相同的游戏方式。过去，类似于芭比娃娃和摩比世界的玩具也曾在许多国家风靡一时。尽管如此，在统一需求和使用规则的同时，这些玩具仍然可以保护儿童的想象力以及他们与传统之间的关系。

最后是产品的使用及废弃处理。我们总是希望在数量和质量上赋予产品更多的性能用途，但是，面对高科技产品，普通用户往往只能使用其中的少部分功能。对于技术产品的维修，通常并不包括基本的零部件更换，一般只考虑维修复杂的零部件，或直接进行升级处理，那么，老旧产品就会被进化版本替代。这种处理方式正在形成一个重大的生态问题，同时也过度消耗了地球的自然资源。如果我们不能够重组生产，节约原料，限制能源消耗，回收处理废弃产品，使用可持续生产方式等，那么地球将无法继续维持。这让我相信，一个真正的文化的革命需要考虑到全球化的各个方面（经济、文化、政治、制度），而不是作为一个不平衡、不平等的因素，应当充分尊重人权，并促进历史和文化和谐发展。同样，我也希望全世界都能思考未来的所有可能的极限。

§
[2]
p.53

伦理与美学

工具箱
摘自《课的语言》，2014 年讲座 3

疑问
摘自《课程、阅读与文字》，2014 年讲座 2

设计师的培养
摘自《设计无处不在》，2014 年讲座 1

智能文化之争
摘自《可期的智能》，2014 年《DIID》第 58 期

homo faber
摘自《真假辩论》，2014 年《DIID》第 57 期

演出
摘自《突变》，2010 年《DIID》第 44 期

陷入混乱
摘自《过百设计师》，2007 年《DIID》第 33—34—35—36 期

1 个 H，4 个 W
摘自《不同 & 设计》，2006 年《DIID》第 24—25 期

细微或显著
摘自《不同 & 设计》，2006 年《DIID》第 24—25 期

理由
摘自《量产设计或无名物的力量》，2005 年《DIID》第 15 期

高科技美学
摘自《高科技》，2004 年《DIID》第 9 期

系统共存
摘自《伦理 & 技术》，2003 年《DIID》第 6 期

伦理 & 思想
摘自《全球设计与附加伦理》，2003 年《DIID》第 3—4 期

主导品位
摘自《精简艺术》，2002 年《DIID》第 2 期

对许多人而言，尤其是从事设计专业的人来说，伦理学与美学仿佛是一枚硬币的两面。从广义上来说，伦理学涉及与政治、道德价值相关的人类行为模式。而美学通常以追求美为目的，感知我们所处环境之美。伦理学是一种意识形态，而美学是一种体验。伦理学让我们能够支持美学，而美学又是其进展中伦理的必要条件。无论是设计的概念还是物质形态方面，如果我们没有思考清楚如何在其生命周期中将"伦理学"和"美学"概念结合，我们将无法真正设计。这个愿望在意识形态上似乎是完整的，但是在当今的设计实践中却越来越少了。

§ [2] 设计 _ 文本与语境

工具箱 > 我们清楚地知道在设计领域，图像即文字。因此，图像像文字一般拥有绝对的语言自主性。然而，即使图像具备良好的沟通能力，我们也寄希望于完整的、具体的语言责任。玛乌立佐法·拉利斯（Maurizio Ferraris）曾明确地赋予语言三个绝对的价值。他认为，语言必须永远是思想的结晶（如果没有真正的思想，那就不存在真正的语言，因为思想是意识的镜子）；语言应当具备务实的可操作性（可以理解为在一定的条件下使用，除了单纯地作为书写工具）；最后，不是所有的事物都可以诉诸语言，因此语言不能言说所有，也正因如此，"文明的危机通常表现为语言的危机。语言可以通往真理，但也可以通往谎言和失去本意"。

存在一种众所周知的说法：语言是一个"工具箱"，其中每个工具都有着各自不同的用途。正如英国哲学家约翰·奥斯汀（John Austin）所表明的一样，"语言还可以用来制作东西，建造某种类型的物品，它的功能并不局限于描述与沟通，而是创造物品"。约翰·奥斯汀（John Austin）的话足以说明，当我们面对科学和方法论方面的事物时，了解它们的复杂性是十分必要的。正如当我们听一堂课时，除了内容以外，还可以借助于主讲者所使用的语言，来验证课程的权威性。

吾苟·沃利（Ugo Volli）说过："人，是会说话的动物，人，生活在语言之中。这一言论成为从柏拉图（Platone）到海德格尔（Heidegger）的所有哲学的基础。今天，我们并没有改变这一最初的想法，但是我们可以将语言纳入一种更加宽泛新颖的范畴，即沟通交流范畴。"也就是说，我们不能忘记"语言即沟通"[正如1964年，马歇尔·麦克卢汉（Marshall McLuhan）观察与现代社会相关的大众媒介时写道，"媒介即信息"]，"尽管如此，我们仍需要谨记，每一种沟通行为都是一种对话行为，而对话总是由不止一个人以民主共享原则为基础产生，因此，它往往涉及道德伦理"。

疑问 > 从这些观察中产生的疑问，质疑我们在设计文化领域的主导意义，质疑这些主动性在知识界内外的作用。简而言之，就是怀疑推动一切发展。对卡迪尔（Cartesio）来说，疑问，是智慧的起源，也是所有问题的起源。

2007年至2008年间，埃托·索特萨斯（Ettore Sottsass）为展览写过一本有关自己的书，其标题为《我想知道为什么》，正好符合我们的观点。在展览过程中，埃托·索特萨斯（Ettore Sottsass）向采访他的人强调，"我面对所有的事物总是持怀疑的立场，我也不知道为什么会发生"，但就是这样发生了。在书后面的几页，他就"先锋"这一概念阐述了自己的看法。他深信这是一个"禽兽"的词汇："对于先锋，我有话要说。当它被这样定义的时候，我不感兴趣。有人认为自己处于前沿位置，我也不感兴趣。成立孟菲斯时，人们问我们要进行什么样的革命，他们视我们为先锋派分子，而我们恰恰不是。我们只是芸芸众生中的一员，我们只希望可以使用新的材料设计新的事物。"我认为，通过埃托·索特萨斯（Ettore Sottsass）对于新事物的描述，我们渴望对周围世俗的事物进行充分的了解，包括它们的缺点以及脆弱性。在这种情况下，设计需要真正了解其作用，并随时做好改变操作工具的准备，因为一切都会随着更小的约束更大的实证而改变。作为学者和教师，我们有机会了解解放设计文化的需求，虽然这些需求是自发形成的、出乎意料的，甚至不一定是正确的。但这是一个不容小觑的现实。

设计师的培养 > 设计，通常是在日常生活的维度上进行的表达。因此，它不仅存在于少数群体的常用物品中，还广泛应用于大规模的消费产品中。它涉及所有的生产部门及企业部门，并清楚地表达了先进工业化社会的设计需求。

设计，其传统的任务是赋予产品外形，而如今涉及的是整个工业设计的复杂性。因此，设计师的养成需要长期且系统的学习。接纳先进的知识，理解供需关系；认识并了解应对复杂性的工具与方式；提供工具来分析需求，不论是在性能还是在生产技术方面，并将其转化为创新物品。

智能文化之争 > 这就是为什么那些从事设计的人必须清楚地认识到，想要将智能概念引入产品，首先需要具备对"智能产品"有着集个人、理论、方法论、智能产品功能于一体的综合认识。设计师必须明白，新的科学知识促进发展新的技术以改变或生成新的材料，同时也能够改变生产方式，甚至改变了产品转化的原因。难以想象的是，将一些物品稍作修改并借助无线设备赋予它们新的生命，就称之为智能产品，随之开启的是智能文化之争。此外，将新技术加入并补救已存在的标准类型，来丰富产品及产品体系功能，也是不合适的。至今为止，也许唯一没有违背"智能哲学"的真正智慧就是材料本身，在设计师实验者，乃至进步公司的眼中，它们才是智能创新的根本。举一个纹理领域的简单例子：受到许多偏见的fòrmica（塑料层压板的专利名称），如果不能够模仿木材的纹理特质，那么它将无法立即取代之前的材料并投入使用（正如罗兰·巴特一开始并不信任塑料这一材质，却也在后来的《神话修辞术》中对其进行颂扬）。fòrmica（塑料层压板的专利名称）或者直接说塑料，经过了几年时间的发展，它们不需要再进行模仿，而是通过自己的特质变得流行起来。一旦它们的结构构成和可操作性得以完善，那么，塑料就能够将新颖的色彩与缤纷的感觉带入家庭环境。这种情况下，对于材料的使用，总是存在两种方式：纯粹及虚假。前者让材料具有新的更加自由的语言，而后者则利用材料自身给人造成错觉感。

homo faber > 随着消费社会经济和政治的改变，设计师形象也发生了改变。对这一观点的实质争论，是由一系列价值方式直接造成的。开放源代码、开放性设计、社会设计、协同设计，它们之间的自给自足，超越了艺术想象及个人主义维度，重新恢复了现代设计的公共模式。这种方式赋予设计新的基础，成为跨越个人才能、集体才能、新的使用及美学范式之间的桥梁，使得如今的设计具有实用社会的文化支撑，可以从各个角度各种状态下得到大众的理解及接受。与此同时，设计师个人主导负责整个设计方案这一方式也依然存在。因此，如果想要在这两种不同立场之间找到一丝联系，我们应当期待现代homo faber在手工制作（无论是使用手工工具还是电子工具）原则的鼓励下再次出现。所谓homo faber，在哲学机械家马修·克劳福德（Matthew Crawford）的假设中，是指在职业生涯面临解体，所有技能发生颠覆，新旧能力产生冲突的时候，依旧能将工作视为"灵魂之药"的人。在此之前，理查德·桑内特（Richard Sennet）就曾谈论过这个问题。他认为，手工艺人必须同时掌握传统与现代知识，具备前所未有的灵活思想及创造力。社会发展太快，技能随时会被淘汰，工匠们只有通过自我的职业发展来解放个人的才能，才能更好地立足于社会。

演出 > 我十分荣幸地参观了由建筑师扎哈·哈迪德（Zaha Hadid）设计完成的新的MAXXI博物馆。开幕式那天，MAXXI汇集了两万多名观众，这已然成为了一大超越建筑本身的盛事。与此同时，由欧蒂娜·戴克（Odile Decq）设计完成扩建的罗马当代艺术博物馆MACRO也迎来了开幕典礼。

这是一个亲眼见证变化的机会。两个伟大的建筑作品变身为两大临时装置，幻化为一场演出。

建筑物本身的根本特点是其结构性，但它们却被视为短暂的表演场所，完全得益于在建筑内围绕艺术品移动的人和物。

在这所被称为21世纪的艺术博物馆中，伊拉克裔英国女建筑师利用坡道和长廊，以一种流动的方式引领着游客的行进轨迹。游客本身亦成为空间的主角，他们的流动填满了每一个视觉角落，同时，也让他们失去了时间和空间的指向性。建筑本身应当是它们面向全城乃至全世界开放的庆典。但是此时建筑物本身的结构与混凝土，墙壁与屋顶，建造规则与比例，似乎已经虚化成为了背景。而散落在空间里的观光群众，却成为了建筑物中最闪亮的主角，化为环境与美学中的重要元素。

在同一座城市，MAXXI和MACRO，这两个如此遥远又不尽相同的博物馆，它们想要表达的重点并不是建筑物本身，而是如何将复杂的动态转化为一种媒体现象：建筑世界里的两大新兴人物，扎哈·哈迪德（Zaha Hadid）和欧蒂娜·戴克（Odile Decq）的呈现，以及一系列临时展览举办的演出。比如，《台伯河邂逅》，作曲家兼主唱丽莎·比尔拉瓦（Lisa Bielawa）与罗伯特·哈蒙德（Robert Hammond）联合举办了通过展出的100把椅子的城市即兴演出。比如，基罗·得·多米里奇斯（Gino De Dominicis）的巨型骷髅雕塑作品Calamita Cosmica，它经典的长鼻子充满了讽刺的意味，在MAXXI入口处迎接着我们的到来。比如，在MACRO，人们踏上透明的高科技走道，穿过色彩控制的红黑空间。比如，在MAXXI，人们轻轻地穿过连续结构，悠然地漫步在流动空间中。

丰富的作品，众多的艺术家与评论家，大量的音乐与视频，一些好奇的游客，繁多的社交活动与媒体，这一系列的有生命和无生命的事物，这些所有的动态，汇聚成一场伟大而精彩的演出。而MACRO和MAXXI也化身为舞台的结构与背景，成为其中重要的一部分，并遍及整个城市。

观看这场演出，仿佛置身于整件艺术作品前。作为大型舞台装置的主角，MACRO和MAXXI在表演当中及周边尽情地呼吸、感受和表达自己。虽然它们在品质空间的呈现上完全不同，对色彩、语言和技术的使用也天差地别。但是唯一相同的是，在伟大的演出之后，这两件艺术作品最终回归为简单的建筑，观赏它们就好似对比两件艺术作品。

尽管背离了传统的罗马建筑风格，但是新的建筑依然具有吸引力和表现力，随着时间的推移，它们将成为城市发展的固定地标，这与涉及的动态活动的功能无关。

第二次参观 MAXXI 时，那天并非节假日，所以并不那么拥挤，因此，我可以重新解读建筑形式和艺术形式之间的联系。这次经历，让我对它的临时性和演出性有了更深刻的印象。建筑物内部可以加入环境装置、多媒体技术、视频投影等，能够更好地呈现那些不同于简单图形表达的艺术。相对于"容器"，"内容"再一次发挥了核心作用，每次都以灵活的方式塑造出不同的故事。因此，我无法想象当建筑物与其内容分开之后应当如何解读它的意义。如若没有了入口柱廊下横躺着的基罗·得·多米里奇斯（Gino De Dominicis）的雕塑作品，马里奥·梅尔茨（Mario Merz）的圆顶建筑，展示卡洛·斯卡帕（Carlo Scarpa）建筑作品的屏幕，Diller&Scofidio+Renfro 工作室设计的装置的噪音，那么 MAXXI 会失去很多色彩。

在一期《DIID》杂志中描述的大量设计作品与这一新的建筑有着紧密的联系。在此，建筑设计即为空阔空间的设计，试图将表演性及舞台性这两大特征融入其中。

以 Diller&Scofidio 工作室改造的 High Line 作品为例。在曼哈顿，他们将老旧的架高铁路变身为大众绿色人行道。在这个作品中，设计师考虑了一系列环境因素，它们甚至会随着时间的改变而改变作品的外貌。事实上，High Line 作为大众通道，不只有一种形式，它的外形一直处于变化当中。无论是上面的人群熙熙攘攘，还是绿植环绕、四季变迁，都可以改变它的外貌。夏天绿绿葱葱，冬天白雪皑皑，甚至连公共设施都能赋予其不一样的感受。

设计师使用有机材料（自然、植物以及观赏者）和无机材料（经典的建筑材料，如石头、水泥和钢材）所形成的"有机建筑"（agri-tettura，agricoltura-archiettura），再次证明了空旷空间随着生命变化而形成的舞台效果。

艺术赋予空间意义，有时还带有伦理价值。

陷入混乱 > 将如此多的设计师，如此丰富且代表了当前设计灵魂的经验紧密联系起来，这也许就是所谓的混乱。

大量产品侵入我们的日常生活产生混乱，难道我们不正是生活在这种混乱中吗？

日常用品表现形式的多样性产生的视觉污染，这难道不是一种混乱吗？

科技产品的性能过剩造成了一定的操作难度，这难道不是一种混乱吗？

设计师的世界也是混乱的，他们不得不面对设计的各个方面。从有形产品或无形产品，到家用产品或人体产品，从休闲空间产品、办公用品、玩具、机动产品，到工业量产产品或手工定制产品。

在这种混乱中，通过设计师和产品，设计表达了一种渗透在人造环境中的伦理与美学的价值体系，同时也影响了社会行为，统一了大众品味、产品周期和形象风格，并影响了消费者对使用产品的选择。这种情况看似丰富，实际上却杂乱无章，它正是人类学突变的结果，不仅影响了西方社会，甚至波及全世界。就好像是多种风格在不同的地理和历史背景中被连根拔起，碎片散落各地，最后被重新组合形成了一种涵盖多种风格语言的混合体。

因此，将所有的全球传统文化的形象表达结合起来，去除差异，并以统一同化的方式进行表达，这就是全球风格。也就是说，传统文化形式的多样性已不复存在，取而代之的是一种统一的全球化的表现形式。混乱的定义则是全球风格。全球风格展示了以文化及生产关系为背景的当代情景，其中，传统的比重在时间与空间概念上已经逐渐模糊。全球风格的表现形式不再受到物理环境影响，而是更加横向多元化，更加凸显其社会角色，并展示了新的"种族"。总而言之，全球风格是一种产品风格形式，它借助于全球的不同文化与传统的混合语言来进行表现。

1个H，4个W＞消费者似乎对购买个性化产品的要求越来越高，但事实却好像往相反的方向发展。他们试图在购买易识别身份产品与彰显个性产品之间寻找一种平衡。

大型企业用大规模定制生产来响应个性化产品需求。工业化生产过程让所有消费者可以更加轻松地购买个性化产品，即使这些量产产品拥有相同的外观，但是同时也可以具备不同的使用功能。

为了实现这一目标，大型企业开始使用不同的生产方式与技术流程。

为了评估哪些产品零部件更容易实现个性化，甚至可以直接由消费者自己完成，我们需要考虑一些问题：人工造物中哪些产品可以实现个性化（手机壳）？如何实现（通过配件来修改产品外观）？由谁完成（用户本人还是应用户要求的制造商）？在哪里完成（在工厂里、经销商店铺中、用户家中、大街上）？什么时候完成（购买时一劳永逸地完成，还是某些特殊时机）？为什么（为了识别产品还是让产品被识别）？

为此，企业设计并生产了一系列标准零部件，以不同的方式进行组装。通过所谓的延期模式，在生产之后通过不同的方法完成个性化设置。比如复制、替代、定制模块，并将其合并、使用、自由组合。因此，每一件产品，即使是由标准化生产制造完成，也可以实现不同形式的个性化设置。

对于大规模定制，企业需要有能力利用个人用户电子数据库，与其他用户的信息相比较，并通过应用程序，将所有的交互转化为对应流程。

以这样的方式，企业不仅能够察觉用户的潜在需求并迅速响应，同时还可以通过吸取过去的经验，不需要重塑整个过程，就为其他用户提供类似的服务。这也就是简单定制与大规模定制的区别。储存单一用户的需求以及对应程序，以便在遇到相似需求时可以用于第三方。此外，这样还能够让用户共同参与设计，借此减少所谓的"消费者牺牲"，这是大规模定制最值得探究的元素之一。而当用户购买标准化而非定制产品时，则必须承担这种牺牲来强制适应产品，因为标准化产品只是满足普通消费者的一般需求。

细微或显著 > 差异性与一致性这两个概念，在设计文化当中，具有战略及完善的作用。

以建筑设计为例。一方面，建筑物本身是系统的一部分（城市环境、人为环境、地理系统，总而言之就是指一种特定的某个地方的风气或特色）。另一方面，系统参照准则允许建筑物与其所属的整体环境中存在"差异"，在任何情况下都可被识别。因此，可以通过两大元素来形成建筑的基本特征：身份特性及可识别性。

身份特性，指的是建筑作品都是以行业准则（比如建筑外形、类型，所用材料和建造技术，相关尺寸限制以及城市与装饰法则等）为基础，遵守相同之处，分享不同特征，这些是决定参照系统归属感的因素。

可识别性，指的是建筑物通过自己的独有特征与同系统里的其它建筑进行区分。区分的元素可以"细微"，也就是说，即使这一建筑作品具有可识别特性，但同时也与其它同质城市结构特征保持一致。又或者十分"显著"，也就是说，这一建筑作品在参照的建筑结构中明显突兀（在规模、功能、设计语言，以及特殊建筑价值等方面）。

理由 > 在缺乏需求的情况下，大量生产的产品依旧是允许我们在消费社会中重建大众消费的理由。其中，衡量个人的需求以及判断个人消费的标准，逐渐成为可以分享共有经验的关系系统中的一部分。

高科技美学 > 在产品中存在这样一项技术，它能够丰富产品的特性、功能，让其物质材料更加坚固，产品更加耐用。但同时也存在另一项表象技术，它的过度使用使产品变得更加复杂，也引发了对未来在产品及服务方面的更多可能性的探索。

在高科技美学领域，我们正处于这种情况。经常出现过度浮夸的解决方案，除了展现出肌肉组织以外，并无其它。产品使用新型材料，变得更加轻、更加透明，但同时不可避免的是因物品的风格语言存在的时间非常短暂，其性能会迅速过时。

那么，对新产品的设计与生产进行深入研究显得更为有趣。在硬件方面，研究生产过程中技术的应用，而在软件方面，则与高科技美学相关。

无论如何，我们需要对高科技语言的普及进行思考。因为这是充分了解近期多样艺术设计展（建筑、设计、绘画、影像艺术等）的关键。其中，当代性的概念已经多次通过技术本身的救赎理念进行表达，就工业主题而言，它与产品技术的意识形态无关，而与产品的伦理与美学持辩证关系。

高科技美学产生了不少杰作，尤其在建筑领域。有些建筑作品在机构上大胆采用桁架、支索、可见网形元件，但是这些却无法防范火灾，因此需要更加复杂的措施来预防。有些建筑作品使用奢华的轻薄透明的楼板，使得眩晕患者无法进入其中。有些建筑作品配有裸露在外的电机设备或网络电缆，它们由金属合金或具有光泽的材料制成，但是清洁和维护将成为大问题。

高科技美学让建筑物成为一件巨大的物品。在巨大的、蜿蜒的、拟人的建筑外表下,包裹着不同的功能区域,这些功能分区并不涉及指定的活动,也不看重规模比例和历史标准参照。好比日常物品形式特征和技术逐渐转移而来的"大都会怪物"效应。

高科技美学使用幕墙覆盖建筑外观。幕墙使用 LED 和液晶技术,可以通过调节改变幕墙的色彩和亮度,但简单的电路故障足以让它变得毫无生气。

总而言之,这就是一个恶性循环。增大建筑作品的技术使用就是缩短它的寿命周期,增加建筑作品的功能特性就是加速它的老旧过时。

在日用品形式、功能和材料的使用中,高技术美学所表达的,相对于技术的真实潜力,更能代表技术的象征性价值;相对于实验应用技术,更能代表创新思维。但事实上,实验应用技术才能使得产品的寿命更长,产品的最终功能更加合格有效,而没有多余的内容或性能。正如技术在设备小型化中所体现的一样,电话的数字按键或屏幕对于近视或远视患者而言,相对难以操作或阅读。又或者,当技术产品具备大量繁多的功能时,用户却不能完全理解并使用。

系统共存 > "民族与技术"这一主题凸显了全球化生产体系的特点,其中包括了技术的使用及产品的转移趋势。目前,产品逐渐由落后或发展中国家,转向西方工业化国家。具有明显民族特征的大部分日常用品常常被用在时尚、室内设计、通信和技术领域。总而言之,这些产品不仅存在于家庭生活场景中,还存在于工作休闲时光中。也许,如果要深入研究全球化生产这一主题,我们应当从占有大量市场的产品出发。如今我们依旧坚持系统共存,由手工、机械和电子信息化设备共同进行生产。在电子信息化生产领域中,人类的存在感被压缩至最小,优先由机器人来操作器械。同时,在手工制作过程中,即使具有相同的种类和外形,也可以让产品独特性变得可能。

伦理 & 思想 > 什么样的伦理学,什么样的美学,规范了我们周边的人工造物?伦理学与工业制品的质量和数量息息相关。有形和无形的产品进入我们的身体,渗透进我们的周围,甚至蔓延至人类创造的无边际的环境。然而,这些产品涉及过多的商品类别,已然超出我们可能接纳的极限。

科学知识、技术及生产能力不再简单地表现为干预介入日常用品和技术产品,还包括对生物体的操纵。工业产品受专利保护,但如今专利申请不再独属于工业产品,它涵盖领域广泛,包括无生命产品、机械产品、虚拟产品、综合产品、软件、药品、植物品种、微生物、基因、动物,甚至是人类的细胞和蛋白质。美学所表达的语言,是由世界上多种多样的文化和传统的风格混合而成。慢慢地,抵达遥远的地方已不再困难,不同的表达风格和形象化的交替错杂产生新的混合语言。时尚领域的快销产品更迭变换,涉及的风格语言多变且不断有新的思潮涌出,因此,变得更加丰富繁杂。然而,通用语言综合简化形成了一种认同文化。在此，工业产品、日常用品描绘着我们的生活场景，它们及其涉及的市场占据着全球空间。什么样的伦理道德反应了涉及用途、传统、语言、材料、地区生产方式的全球村范围的认同文化？哪些伦理观念让耐克最新系列的运动鞋变得不可或缺？面对相同功能的产品，又是哪些伦理观念让我们优先选择价格更高的知名品牌？药品的流通又涉及哪些伦理规范呢，与工业化市场或第三第四世界国家的规则是一样的吗？又是什么规范着对通信和网络产品的使用和监管呢？哪些价值元素推动创新技术的使用，使得技术产品相对用户的需求和使用能力而言，具有过剩的性能？以全球化为基础，技术能够恢复空间的变量，进而控制时间的管理。正如，利用 autocad 软件的通用语言及网络途径，技术绘图失去个人色彩，在世界上任一角落都可以实现产品的设计与操作，全天候 24 小时无缝进行，而各地区员工只需要在白天继续工作。

主导品位 > 每天，我们都会遇到大量的产品，它们形成了我们日常生活中的各个场景。我们与这些技术产品相互作用，它们构成了我们"物质文化"的标志。这些日常用品的共同特征是拥有不同的风格形式。繁星般的产品有着众多的语言，却没有对应的统一风格。因此，我们缺乏一种"主导品味"，一种主流语言形式或与之相反的研究方向。

但是，多种不同语言的共存正是"主导品味"的一大特征。这种情况看似丰富，实际上却杂乱无章，它正是人类学突变的结果，不仅影响了西方社会，还波及了全世界。

就好像是多种风格在不同的地理和历史背景中被连根拔起，碎片散落各地，最后被重新组合形成了一种涵盖多种风格语言的混合体。这是一种新的国际的风格：全球风格。而国际风格是现代文化运动的结晶，是在建筑、绘画、雕塑、应用艺术上表达的艺术体验。这些前卫艺术不关乎语境（摩天大楼位于巴格达或芝加哥是是同样的），奠定了全球化理论、方法和语言的推广，同时也确定了建筑、功能、风格和技术的统一原则。

相反，全球风格表达了我们这个时代的相对论条件，也就是表现形式的批量化。这些表现形式如果不通过一个过程进行混合，那么，其中的任何一种都无法进行识别。

这种多种风格共存的普遍现象，我们称之为地球村，也就是全球范围内对区域文化的接纳和认同。就丰富的技术与性能而言，有能力展示最多最好的产品的先进工业社会，依旧占有优势。

但另一方面，产品所表现的文化语言却和欧洲国家相距甚远，无论是在图纸上、色彩上，还是材质上，我们都能找到明显的民族文化印记。以欧盟为中心的文化概念步入危机，正逐渐被新兴国家的传统和文化所取代。这些国家的发展潜力预示着它们在未来全球平衡中的重要性，以及科技发展前景中的主导作用。

当我们置身于众多困难中，为了众多零散的语言找寻理论参照，我想，分析新的全球化风格显得很有必要。尤其是当代人工造物的设计和生产的过程涉及众多形象化的表达方式和理论基础，我认为我们应当仔细研究以简洁为导向的方法。

也就是，少即是多。它是上世纪重要阶段之一的典型原则，借此实现了从形象文化的国际化向当代文化的全球化这一重大转变。特别是在复杂且多变的现实场景中，现代化运动的理性主义背后的精神和动力，是极简主义发展的根源。通过它的全球化系统表达，极简主义成为重要的运动之一，我们乐于分析它，因为它是当代风格中最西方的，最具意识形态的。也因为极简主义伴随着简化的伦理思想，这恰恰与我们身边大多有形或无形的人工造物的表现逻辑相反。

简化的美学不是简单地表现在产品的性能和技术的复杂性上，相反，产品正在以新的更丰富的用途，来满足人类日常生活的形式感，来更深入地升华除纯粹美学之外的复杂性，也就是在创造过程中不断地进行理性控制。简化的美学是一种能力，可以从多余功能中选择实用，可以深入了解自然本质，可以在直观 (esprit de finesse) 与演绎 (esprit de géométrie) 之间找到平衡。

自 80 年代中期，"简化" 与 "非物质化" 一直是意大利工业产品研究和实验的主题，并以 "以少致多" 为基础发展了当时十分重要的设计方法。

这是一个不同寻常的方法，无论是与现代运动时期的欧洲文化的功能主义和理性主义相比，还是和 70 年代初期基于紧缩、匮乏直至破坏产品的意识形态发展而来的激进设计相比。

§
[2]
p.63

有用与无用

终身教育
摘自《值得做的，也是值得说的》，2014 年讲座 2
记忆
摘自《课程、阅读与文字》，2014 年讲座 2
建筑的危机
摘自《设计无处不在》，2014 年讲座 1
可整合设计
摘自《可期的智能》，2014 年《DIID》第 58 期
新的平衡
摘自《真假辩论》，2014 年《DIID》第 57 期
大型农场
摘自《从奥威尔到迪斯尼》，2011 年《DIID》第 48 期
与人互联
摘自《生态设计与新愿景》，2009 年《DIID》第 41 期
实践
摘自《过百设计师》，2007 年《DIID》第 33—34—35—36 期
食品设计
摘自《食品工业设计》，2006 年《DIID》第 19 期
认知行程
摘自《材料的真实性与模糊性》，2005 年《DIID》第 13 期
稀有且独特
摘自《设计与奢华》，2004 年《DIID》第 8 期
一只章鱼
摘自《Tool Toy》，2002 年《DIID》第 1 期

讨论"有用"或"无用"对设计师的工作来说是致命的。有时候从概念的产生到结束，并没有诞生任何设计作品，因此在设计领域进行调研是十分必要的。在很多情况下，设计师往往都不能确定自己的首要任务，"有用"和"无用"之间的界限十分模糊，这也是设计院校应承担的文化责任。显然，除此之外也不能缺少想象力，这是极其有趣的游戏，一件产品绝不能将有用作为唯一的理由，我们也需要选择合适的时机传递这些。

§
[2] 设计 _ 文本与语境

终身教育 > 毫无疑问，大学的主要任务是通过提供高等教育和从事实践研究来影响科学文化的社会价值。难以预料的是，大学从事的活动是否能够将其实质内容与抽象意义置于同一层面。之所以难以预料，是因为它始终是一场博弈，它试图通过学科之间的多元文化来营造辩论的良好氛围。所谓辩论，可以促进相互之间的竞争，而不是单纯的学术间的争论，以避免学生被学分教育所限制，推动学生跳出其专业界限来进行知识的相互交流。为此，我们不能信任瑞典的卡特琳娜女王当时的思想言论。她的书籍在市场摊位上被发现，事实上，早在17世纪，她的思想就可以用一句格言来表达，"在学校里学会的东西，之后也会被忘记"。仔细想想，这句话并不完全错误。学校（大学）教育你，赋予你一种自我反思的能力，让你习惯去遗忘某些概念以便腾出新的空间去学习。就好比我们并不提供界限分明的专业知识，而是致力于将各学科知识融会贯通，来对抗那些由学科界限带来的一无所知。

借助于齐格蒙特·鲍曼（Zygmunt Bauman）发表的言论，我们能够更好地理解所谓的一无所知。可以这么说，齐格蒙特·鲍曼（Zygmunt Bauman）从他在《流动的现代性》中，将流动的现代性视为不确定性和变化性的代名词起，就为过度的"预言综合症"所诟病。尽管如此，齐格蒙特·鲍曼（Zygmunt Bauman）在《流动的生活》结尾提出的观点仍然值得一提。他说，"无知导致意志瘫痪。如果我们不知道未来会发生什么，那么我们就无法规避风险。"相对于长期以来以事实为基础进行的详尽考察，面对问题共同探讨解决方案的政权，通过有意识地培养无知和不确定性，进而取得更加可靠的统治并且获得更低的操作成本。政治无知持续不断，并与无知和无所作为交织在一起共同扼杀了民主的声音。为了拥有选择的机会，我们需要进行终身学习。但是，我们更需要维护那些能够提供选择且带来更多可能性的状态。让齐格蒙特·鲍曼（Zygmunt Bauman）如鱼得水的政治层面，事实上是离我们并不遥远的一个层面。因为也许我们最缺乏的是政治深度的意识，缺少这种意识让我们的工作变得薄弱且毫无效率。这种意识可以让我们更好地领悟：怎么才能让优质大学得以存活，能够知道如何且何时赢得大众的关注。

记忆 > 我们在文化意识方面的表现是一种超越了自我认知的相互比较。在时兴参与政治的时候，在辩论概念转移到我们的专业领域之后，"比较"一词便得以释放。有多少次我们使用"建筑中的辩论"，"有关城市的辩论"，甚至"有关环境的辩论"来进行表达？而最近几年，"辩论"已经转向人工产品领域，全球化与本土化之间的比较似乎毫无意义。显然，我们是善辩的人群，在学术领域，我们往往是预见者，有时是煽动者，但更多时候是人群中慌乱的执行者。

因此，我不得不相信，只要我们的认知受到质疑，我们就应当立即求助专业领域的权威人士。他们的这种权威性，来源于大量的研究，来源于知识传播原则的价值观，来源于教师和研究者担负起的真正的社会文化职责的使命。事实上，这些都是我们熟知的事实，但是时常被我们遗忘在脑后。

建筑的危机 > 城市场景的变化正持续不断地从建筑转向设计，这也表明我们的城市不仅在建筑的常规

有用与无用

方向上发生着改变，而且通过产品的扩张加速了这一变化的进程。城市的景观，视觉思维的潮流，科学技术甚至生活方式的美学，都将取决于设计的演变，也促使了其产品、服务和策略的多样性的变化。

对于以维多利欧·格里高蒂（Vittorio Gregotti）为代表的许多激进建筑师而言，设计的"成功"表明建筑已经陷入危机，尤其是在语言、风格，解构能力方面。维多利欧·格里高蒂（Vittorio Gregotti）认为，设计涉及了众多专业领域的知识，因此造成了一种思维上的混乱，使得有着望远镜一般外观的建筑物得以存在。此外，因为设计学科对市场和生产部门进行的"妥协"，破坏了对建筑领域的投资和目标之间的关系，尤其是将时间因素和废弃原则引入建筑设计的时候。正因如此，设计是当代城市运营的基本推动力之一。试想一下，城市借助设计在其中心重新赋予废弃空间以功能性，并以"技术和技术设备的试验生产者"的身份，通过持续不断的设计作品，来进行自身发展以满足社会需求。这一过程是自发形成的，也是难以预测结果的，在这种情况下，持续不断的设计作品以一种不太恰当的方式填满这些空间，从而赋予职业以新的契机。

相反，在艺术与设计之间，存在一种绝对清晰的状态——我中有你你中有我。随着时间的推移，艺术与设计消除了一切障碍，找到了共通的空间。然而，描述这些相同之处并不意味着设计产品与展示的艺术作品一致（这种情况与其它因素有关），也不表示大量展出的艺术作品与设计产品相同（从马赛尔·杜尚和曼·雷的 Ready Made 现成品艺术开始）。而是在二十世纪的理论当中，艺术所提出的相关概念，设计早已以某种方式对其进行传播与推广。

可整合设计 > 不同的主题和文字始终贯穿于理论现象与具体设计当中，这是极其自然的现象。它们不断地激发设计的吸引力，并决定了其结果。长期以来，我一直与"现代性"的定义做斗争，而如今，当畅想未来的各种场景时，我们试图在城市、家庭中与所有事物进行斗争，"智能"这一概念又将我压倒。作为夸罗尼（Quaroni）学院毕业的设计师，一直以来我奉行先理解后执行的原则。我认为在广泛传播的设计文化当中，变化的内容并不需要过多的口号。几年前，我们所追求的整合设计，是融合各方面的知识来确保智能事物的成功。而如今，我们试图研究周边的人工智能，但在高科技面前，依旧会感到渺小无力。在"智能"这一关键词中，矛盾也更加具象化。复杂的"设计过程"赋予"变化"以形式，而此时，"智能"似乎成为唯一能够阐述这一过程的关键词语。很长一段时间，我思考"智能"与"建筑"之间的关系，即使我感受到"智能"的广泛含义，我也不禁要考虑由于过度使用而带来的隐患。之所以称为隐患，是因为它看似完全无害，为了消除对未来的无法预见的焦虑，我们应当避免将它们视为真正的设计探索者。对于城市而言，"智能"代表着更加实用、紧密相联并包含各种在线服务（智能手机能够完全覆盖）的想法。然而，这种以新的通信技术为核心的良性城市系统，似乎不足以提供一个以 forma urbis 为概念的新的原型。总之，就所谓的民众的互联网互动而言，在不具备合适的解决方案之前，无法有效地将过去的有形或无形的作品与现今的作品进行区分。我们采用工业产品领域的类似想法，缩小设计尺度。虽然我们十分擅长将智能物品（产品具备互动、行为控制、记忆、加热、制冷、刺激感官等性能）进行分类，但在分析它们的美学及伦理尺度方面却有所欠缺。我们正处于设计学科的关键阶段。我们观察着一个正在发生的现象，它多变且难以遏制。但同时，我们也需要思考产品设计涉及的所有概念，从产品的构思到外形结构，等等。

新的平衡 > 那些了解我学术兴趣、意识形态立场的人，知道我面对"工艺"这个词时的不安，有时甚至是本能的不安。因此，我不接受，也不能接受以这种传统的、工具化的甚至是因循守旧的手工文化，来定义设计的起源（并不是工业设计）以及阐述设计的未来。无论这一想法是正确还是错误，人们在计量手工艺在工业产品生产中所占比重的同时，却忘记考虑在此期间，工业生产的格局在形式和内容上已经发生了改变。因此，我的愤怒从来不是针对这一典型的意大利"制造文化"。而是那些预言家、大师和手工艺者（如果是艺术家则更加糟糕），他们试图恢复在"纯粹"的手工制作领域的"绝对真理"，事实上，这种"纯粹"是不存在的。相反，正因这些思想的存在，我愿意阐述我对于"Makers"的想法。如今在模棱两可的背景下，随着设计、生产、消费和消费者之间关系的发展，Maker现象的出现，颠覆并改变了所有人的观念。我想说的是，Maker战略宣称，它代表了人工造物世界中供需关系可持续发展的一种新的平衡模式。但这种说法是不恰当的，因为制造业在其传统领域的使命还没有完全结束。如果真的出现这种转变，那么确实会对所谓的生产地区产生冲击，首当其冲的就是发达国家（日本、德国和美国），同时也包括那些接纳先进地区工业离岸外包产品的东欧国家。其次，受影响的还有真正接纳新鲜事物的韩国以及中国、印度。也就是说，城市、住宅、办公室、公共建筑（图书馆、学校、大学、医院、法院、健身及运动中心）将不得不被智能物品填满（只要参加类似柏林的IFA电子展，你就会明白）。设计者众多，生产者亦然。因此，我们不能掉以轻心。在家中使用机器就可以自行生产物品的人，他们以为找到了避开货物流通的方法。事实上，除了罕见人才（如：最近发明可以用盒子包装的汽车底盘的人）以外，大多数人冒着风险希望自己的产品销售于网络市场，但最后却发现它们在货摊之上无人问津。

大型农场 > 《为宠物设计》，这一标题就能说明一切：这期杂志涉及的专题、方案及产品，无不阐述了人类与动物之间永恒且互补的关系。在人类和动物之间存在一种具有所有限制和希望的"大型农场"，如今，设计者已然成为相对其它领域更为道德且更具功能的角色。

设计一个假定的"农场"，我们举两个比较极端的例子。首先，是由乔治·奥威尔（George Orwell）在1947年提出的《动物庄园》。那里的动物厌倦了人类的剥削，它们试图反抗以摆脱控制。但是，随着时间的推移，它们又使用相同的工具与手段，无论是在行为动作上还是外在表现上，模仿着主人的一切。另一个例子是由沃尔特·迪士尼（Walt Disney）提出的假想的《农场城市》。在这里居住着由动物装扮

有用与无用

成的人类，或者人类装扮成的动物。它们积极、怪诞、友好，甚至吹毛求疵。这些童话是我们的文化和想象力的一部分，包含了我们拥有或者期许得到的一切。

因此，如果我们的宠物（我们的动物朋友）像人类一样，这没什么好惊讶的。对于它们来说，它们同样具有甚至超越人类的一切。人类投射在它们身上的不同的情感（激情、狂热、恐惧、娱乐），与集体实践和伦理问题息息相关。而以此形成的设计文化背景，完全呈现于典型的具有当代设计特色的人造机械设计中。

我们研究各种适用于动物世界的扩展设计，从肉体形式到家庭与环境层面。事实证明，我们分配给时而温顺时而暴躁的动物朋友们的时间、空间以及相关产品，它们具有两大设计主题：通过人类想象并制造的动物生存所必需的物品，来阐述动物所处的环境；借助动物伦理与人类美学相结合的产品，来反对人类与动物共存原则。这种伦理以动物园及主题公园的形式呈现，以娱乐和表演为目的对自然栖息地进行模拟。同时，"宠物时尚"现象日益增长，人们对服装和造型的痴迷也转移到了无辜的动物身上，让它们成为了新兴时尚的受害者。

与人互联 > 我想要描述最近的《巴黎设计突变》展，来分析一段与趋势截然相反的经历。

展览以变革为主题，并提出一系列的实验想法，这显然也包括可持续发展的问题。如果没有这个环保问题，则不会阻止产品的深入革新，使它们不时地呈现出全新的现在和美好的未来。

真正有趣的是，这些公共机构（Délégation aux Art Plastiques/Bourse Fiacre，Buorse AGORA cuor le design，CulturesFrance，VIA-Valorisation de l'Innovation dans l'Ameublement）开始通过询问设计师们，甚至是非常年轻的设计师，来了解我们的未来发展方向。

让设计来阐述我们生活的变化，在新场景中通过设计提出有用的产品或用产品体系改善人类的生活。这对于一个国家来说，是非常有远见的。

为了革新居住文化，让·路易·弗里山（Jean-Louis Fréchin）潜心研究了电子技术，他在简化产品的技术设备的同时尽可能地发挥其潜力。比如，WaSnake是一个架子，但同时也是一个通讯设备；WaazAl是一个架子，但同时也是一个扬声器；WaDoor不仅是一扇门，它也是一台幻灯机。

让·路易·弗里山（Jean-Louis Fréchin）利用新型产品，大大减少了居住空间中产品的数量，并为它们与人类的相互联系做好准备。以这样的方式，他提出了一种全新的家庭环境，在此，所有日常物品可以与网络和家庭电脑连接起来。他是数字革命的代表人物，并将数字技术视为赋予无形经济以身份形式的契机，他借此改变了人与物之间、人与时空之间的关系。

弗朗西斯·阿扎布赫（François Azambourg）采用近乎手工的方式将材料的性能运用到极致。无论是织物或是木材，金属或是树脂，胶水或是泡沫，它们的所有能量都能被完全释放出来。因此，这些物品看似是风格的运用，但是仔细观察之后却发现它们所表达的是物质经济伦理所引导下的设计。正如他布置展出的一整套椅子，通过交织材料的应用，赋予了产品新的外观形式，并运用其美学理念简单直接地讲述了它们生产制造的过程。

在马修·勒阿奈尔（Mathieu Lehanneur）的世界里，我们的身体与生存的环境之间存在着多重联系：温度、声音以及空气。我们自身的存在，我们的平衡系统都受到外部参数的影响，我们不断地进行改变来适应我们的生存空间。因此，他展示的作品促进了居住空间的改变。Bel Air 是一个利用植物的自身属性进行空气过滤的系统。K 是一台光感器，白天接收日光，夜晚在家庭空间中释放光亮。O 是一台氧气制造扩散机，B 是一台音频扬声器，而 C 是一台红外加热器。

EDF R&D 设计工作室研发的产品通过告知使用者日常活动中消耗的能量数量，来借此促进人们生活习惯的改变。如果我们正确设置 Coupe-veille，通过其中的传感器，利用颜色的变化来显示消耗的电量。利用同样的原理，该工作室的研发团队试图设计出能够产生能源的产品。

实践 > 设计师从未放弃自身固有的对独特作品的艺术追求，并试图借此来阐述设计的工作。即对产品的外观形式有着不断的追求，来满足人们的需求，愉悦感官系统，完美诠释产品的美学价值。

这种情况下，设计师通过实验来验证产品的各种可能性。他们试图在运用新型材料、新的生产系统，以提出新的产品类型，这通常都是独一无二的作品，与此同时，他们成为生成的新风格语言的"引领者"。这些作品有些极具创新意识大步迈向未来，有些则是真正的艺术作品见证着当代文化。从这个意义上说，伯特（Humberto）和费纳多·坎帕纳（Fernando Campana）研究使用简单材料、工业废料来设计独特产品，他们的这种有效的表达方式具有特别的意义；通过洛斯·拉古路夫（Ross Lovegrove）的作品可以更好地帮助我们学习了解如何走向未来；罗伯特·卡布奇（Roberto Capucci）的织布雕塑作品，极具代表性；盖特诺·佩斯（Gaetano Pesce）是新材料的实验者，探索生产"不同的系列产品"；三宅一生（Issey Miyake）则不断地探究创新技术。

但是，设计师阐述人类的需求，借助自身的天赋将需求转化为产品，无论是形式还是功能方面，都没有任何多余的部分。学习如何最少程度地使用材料，简化生产步骤，为人类活动提供更多的可能性。从安东尼奥·奇特里奥（Antonio Citterio）的作品就可以看出，他在使用丰富的技术手段的同时，还能使其产品具有精致的极简主义形式；马塔利·克哈塞（Matali Crasset）专注于诠释人类日常生活的仪式感，并为人类新的行为设计了新的产品；马克·萨德勒（Marc Sadler）、布鲁斯·费菲尔德（Bruce Fifield）、雨果·高根（Hugo Kogan）、安迪·戴维（Andy Davey）、米凯莱·德·路奇（Michele De Lucchi），他们将创新技术注入大量产品，其中包括家用电器、医疗器械、诊断器械、照明设备、通信设备，以及运动装备等；布鲁莱克兄弟（Ronan & Erwan Bouroullec）的作品则是以灵活性、多样性、可逆性、可组合性为基础，其中的元件也可以相互结合。

食品设计 > 70 年代后期，法国新式烹饪声名鹊起，为食品设计的兴起创造了新的空间，并统称为

Food design。所谓食品设计，涉及大量文化实验、创新领域的同时，也与传统知识与实践相结合，利用先进知识、技术和逻辑，呈现了自然或人为的食品场景。厨师进入设计领域，而设计师则赋予食品美学及形式价值，他们共同创造形成了以食品为主题的新的领域：菜品如同服务，食物如同美学产品。通过设计食物，结合厨师和设计师的经验，研究实现"餐桌的布置"；通过研究食物与餐盘之间的和谐搭配，试图全方位地刺激人类的感官，带来一种全新的饮食体验。因此，这一领域涉及多元化的知识与能力，除了烹饪以外，还包括人体工程色彩学以及几何图形形态学，等等。

食品设计，除了借助食物的色彩及形式进行各种可能的组合外，还涉足了新的创新领域。以最近保罗·巴里凯拉（Paolo Barrichella）的食品设计工作室在米兰进行的展示为例，在米兰国际家具展期间，他在《城市之光漫步》展览中进行一场名为《可饮之光》的活动。保罗·巴里凯拉（Paolo Barrichella）利用汤力水中的奎宁氯化物的属性，使用金巴利苏打开胃酒作为可饮用灯具材料，他将这个灯具置于在黑暗之中，透过尼龙管射出的 UV 射线，营造了一场灯光视觉盛宴。以这样的方式，就能实现在不使用能源的情况下将液体变为光源，并随着人们自由"行走"。

食品设计，涵盖了广泛的产品设计领域。比如从杯子、碗盘到刀叉的享用食品的器具，从开瓶器到核桃夹的桌上小用品，以及所有的配套的炊具。

认知行程 > 早在半个世纪以前，罗兰·巴特（Roland Barthes）在他的《神话修辞术》（1957 由 Édition du Seuil 在巴黎出版）中，阐述了对材料和物品极其矛盾的情感。这些材料和物品由传统或创新的生产模式形成，而这两种生产活动本身就是相互矛盾的。至今为止，这两种模式涉及的材料与产品之间关系依旧是进行阅读的关键。其中一篇文章对木头或非传统材料制造的玩具展开描述："眼下的玩具都是用让人看了不舒服的材料做成，是化学制品，不是天然物品。他们往往以混合的糊状物模压而成。所用的塑料外表上粗糙但又卫生，减弱了触觉的舒适、柔滑及温厚之感。木质玩具逐渐消失了，这是个让人惊讶而沮丧的现象，就结实而柔软、接触时自然而然的温热之感而言，木头都是理想的材质；它可以从各个面上削去过尖的角度；孩子搬弄它，碰撞它，它不会震动，吱嘎作响，出声却是低沉而清脆。这是种亲切而蕴满诗意的材质，让孩子仿佛与树木与自然接触一般。木头不伤人，也不易被弄坏；它不会破碎，只是被磨损、用旧，可持续用很久，伴随着孩子的成长，渐渐改变外在物品与手的关系；即使它寿命到了，也只是逐渐缩小，并不会呈现膨胀的状态，不像那些机械玩具因弹簧坏了而局部鼓胀起来。木头制成了蕴含物本性的物品，永恒的物品。不过，木质玩具现在几乎绝迹了，它们确实只有在工匠时代才可能存在。往后，玩具在材料和颜色上都将是化学品；材质本身引起的是使用的体感，而非快乐的体感。另外，这种玩具很容易损坏，一旦损坏，就没有丝毫被忆念的生命存在于儿童的心里。"即使罗兰·巴特（Roland Barthes）描述的都是对玩具的感受而非材料，但是相对于所表达出的对木头的热爱，他在几页之后的《塑料》一文中，却对塑料进行了完全不同的解读。"尽管它们的名字让人联想到希腊的牧羊人（聚苯乙烯、酚醛塑料、聚乙烯、聚丙烯），但事实上，塑料的本质是化学物质。最近塑料产品集中展出，展台开端处公众大排长龙试图窥览这神奇的操作；物质的转换；完美的长管形机器（展示一段秘密行程的最佳形式，意味魔术），毫不费力地就从许多浅绿色的水晶中提取出了开了槽的闪亮小碗。一边是天然的原材料，另一边却是完美的产品。在这两端之间，站立着一个带着帽子的员工，他仿佛是半个上帝，又或是半个机器人。以这样的方式，与其说塑料是一种材料，不如说它是一种蕴含自身无穷尽变化的思想。正如它粗俗的名字一般，无处不在也随处可见；正因如此，从另一方面来讲，塑料是一种魔幻般的物质；它的神奇之处在于它对自然进行的粗暴的转化。塑料保留着这整个震动痕迹，因此与其说它是这一过程的产物，不如说它是转化运动的轨迹。"

这两种对于材料的描述，看似矛盾实则在同一路程中建立了两条不同的认知行径。于第一条行径，在木头成为产品材料之前，我们已然熟知它的故事。于第二条行径，塑料在运用过程中隐藏了其本质和复杂的转化性，在视觉及触觉上给人以一种"模糊"的感受。因此，不再是塑料的语言粗俗，木头的语言精细。而是，木头的文化表达了人类纯粹的手工劳作，利用器材进行切割、刨平、打磨及抛光。塑料的文化阐释了一个密封的化学过程，通过人类的介入，魔术般地将其转化生成为一种材料，任其持续发展，即使生产行为完成之后，塑料依旧能够与人类、时间及环境相互产生作用。

稀有且独特 > 一个工业产品是否获得成功，取决于它的普及度和公众满意度。产品的需求越高，生产的数量越多，那么，公司的业绩和利润也随之增长。这不得不让人联想到"奢侈品"，这种"稀有"且"独特"的产品，似乎并不能与大量生产的工业制品相提并论。但事实上并非如此。奢侈品的定义及其表征，都受到媒体交流、市场战略及风俗传统等因素的强烈影响。早前，奢侈品是独一无二的，它们拥有很多价值，比如使用考究及高品质的材料，熟练技工的制作工艺，有时，奢侈品甚至是一件艺术品。而如今，这些价值已然被奢侈品品牌的形象价值所替代。

拥有并展示一件物品，是社会地位的象征。正因如此，我们对拥有一件物品的愿望，容易被媒体工具所影响，从而改变我们的选择。它不仅推动着我们购买一件著名设计作品，甚至驱使我们去收集整套产品。比如市面上琳琅满目的斯沃琪手表，色彩缤纷的鳄鱼服装，最新系列的耐克产品，托德斯的新款鞋子，飞利浦·斯达克（Philippe Stark）的作品，以及具有卡通形象的阿莱西产品。

存在一个相当矛盾的情况，在香港对于发动机的使用有着严格的制度，此时只有通过改装强大的发动机以降低其性能才能适用于香港的街道。但是，法拉利是 F1 赛事中的优胜品牌之一，拥有法拉利，就意味着你拥有着能进入汽车赛道的强大座骑，也意味着你具有他人难以企及的身份地位。因为，法拉利似乎都按需量身定制，它的预定途径和交货时间让其本身更具独特性。事实上，顾客自身也能够感受到法拉利的过高价位相对于投入的创新研究，并不是十分合理。但是，精神的需求同样会衍生出对独一无二的产品的渴望："希望可以通过拥有某些东西，使得我们身心愉悦，享受时光，并自我满足。某些情况下,物质产品的独特性还包括其本身固有的美学外观。虽然这些独特性难以在工业量产的产品中出现,但是,需要我们重新从根本来解读'稀有'和'独特性'的概念,事实上,它应当是一个由设计师、企业以及'智能设计'共同完成的任务。"

一只章鱼 > 如今,相对于产品在文化上的有用功能,多余的外观似乎更占优势。在此,产品形式总是与明确的功能处于相对关系。从而导致在产品设计中存在大量的模糊性与错觉感。无论是形式还是材料,都并没有与最终的使用目的形成一种有机关系。正如,在章鱼的外表下隐藏着一台榨汁机;产品外形给人以一种错觉感,我们以为是一支笔实际上却是一把牙刷。形而上学的模糊的外形,未知的材料,以及无法猜测的大小规模,使得领悟产品的本质难上加难。在这种情况下,我们只能将产品视为一种情感化物品,它与情感的结合试图让所有人与其产生共鸣。因此,此时的产品变身为欲望产品,从各方面刺激人们的感官系统,然而这种趋势早已存在于ToolToy美学历史中。

§
[2]
p.72

混合现实

学术课程
摘自《课程的语言》，2014 年讲座 3
相互影响
摘自《看得更远》，2014 年《DIID》第 45 期
交互设计
摘自《技术依附》，2009 年《DIID》第 39 期
混合过程
摘自《设计中的混合文化》，2008 年《DIID》第 37 期
电影设计
摘自《电影 & 设计》，2007 年《DIID》第 28 期
社会本质
摘自《伦理 & 技术》，2003 年《DIID》第 6 期
转移
摘自《意识转移》，2003 年《DIID》第 5 期
微型设计
摘自《简化的艺术》，2002 年《DIID》第 2 期

科学告诉我们，将特性有所差别或完全不同的生物体结合而成的是一种混合形式。从广义上看，设计从来都是一种混合形式。考虑到它的众多的参与者，涉及的主题、工具以及目标，它的内容是混合的。设计人工造物时所运用的延伸物是混合的。设计天生的后现代倾向也是混合的。混合就好像是一个包含了所有评论与定义的集合，让任何人理解产品成为可能。当设计试图与现实、变化进行较量的时候，它所采用的语言及工具能够适应改变，这也是混合的。

学术课程 > 在哲学家玛乌立佐·法拉利斯（Maurizio Ferraris）的《哲学的问题》的序言中，我读到这样一段话："我们每天都在聊天，也许我们讲的过于多了，有时甚至在讲述过程中都没有进行思考。这似乎不太可取，不仅仅是因为它造成的影响，还因为它破坏了语言的含蓄前提——每个词汇都隐藏着一种思想。为了更好地阐述这个前提，更好地了解其重要性及其限制。我们要深入思考三个重点。第一，有关思想与语言的关系；第二，有关语言的务实尺度，也就是语言除了思考与描述世界以外还能做些什么；第三，最后引用哲学术语中存在的caveat概念（小心谨慎，引以为戒），来对抗olismo liguistico概念（所有都是语言，语言可成为所有）。因为事实不是这样的，有些事物是语言无法阐述的。"我之所以选用lectio（拉丁语中lectio的意思指学术课程，授课者具备很多学识）为标题，是因为我认为我们正处于社会文明发展之中，而此时大学的任务是推动科学文化的社会价值。在其它的教学场所，我借助设计词汇的含义讲述了教师们在选择议题时所面临的责任，他们向学生传递知识时，不应当表现出百分之百的确定，应该传授他们从实践、实验研究和理论思索当中做出质疑的能力。在听过很多讲座之后，我开始认真思考语言的主题，及其作为教授课程工具的本质。

相互影响 > 艺术、建筑和设计之间的独立学科关系已不复存在。它们在操作和表达领域中的学科界限开始相互渗透，相互交融，相互影响，直至一个新的参考体系的建立，使得建筑、设计与艺术之间相互吸取彼此的实践经验，在语言、设计方式和产品形式方面形成了混合。

因此，探究这个新的场景是十分有趣的。其中设计的表达被扩大至艺术领域，艺术也会借助产品来阐述当代性，同时建筑运用设计文化美学，并采用相关材料理论及技术甚至是生产过程。

这一场景集艺术、建筑和设计学科于一身，重新定义了一个全新的、复杂的领域。在威尼斯举办的最新一届由妹岛和世（Kazuyo Sejima）策划的国际建筑双年展上，她以一种有效的手法展示了与设计、艺术相互影响的建筑概念，进一步以一种清晰专业的方式描绘了这个新的世界。

她沿着这条路径讲述故事，形成了对建筑学的形而上学的描述：即相对于建筑的最后呈现，应更加关注建筑作品的构成过程。事实上，这是一种追踪建构材料混合的有效方式。

46位参观者在展览装置周边徘徊，他们越过亚里士多德学派路线，遵循妹岛和世（Kazuyo Sejima）的逻辑思维，对"关于思考"这一主题进行深入挖掘。

威尼斯双年展中，妹岛和世（Kazuyo Sejima）提出的精神和物质理论之间有着共同点，即今日的产品设计不得不面对新的建筑理念，建筑已然成为一个可以容纳产品和艺术美学、伦理价值的场所。追随着展览的这条主线，即使对展出内容有着不同的理解，我们渐渐对建筑种类的分类失去兴趣，开始更加关注建筑的文明化；我们渐渐对雄伟壮观的建筑物失去兴趣，开始更加关注它的轻便性、不稳定性，它与人的交互性，以及在各种场所的适应性。

这也就是满足人们各种感官需求的建筑。它仿佛一件艺术作品，其中的光线、色彩刺激我们的视觉，其中的材质不断地刺激我们的触觉。我们所有的感官系统都被积极调动参与其中,这与建筑传统的功能定义完全不同,进而促进人与建筑作品之间,及其实用性、感觉性、产品交互涉及的技术之间的关系。

无论是处于都市还是郊区,又或是大自然中,这都是一种人与建筑作品之间的全新关系。人类逐渐成为建筑物的核心部分,这在过去是从未有过的。随着观察者的角度改变,建筑物也随之发生变化,它逐渐失去物质感,变得透明变得轻盈;光线照在其中赋予其不断的变化性,使其变得无法触及;它不再参照欧几里得定律,变得活力灵动。

建筑作品具有表达所处地域特色的能力,给人一种熟悉的感觉。与此同时,它仿佛被突然置入,好像一场艺术展览。总而言之,最初我们无法在现有建筑中找到熟知的建筑原型,因此我们迷失其中。而今,我们重新找回方向,我们明白建筑已然成为一种新的当代性的表达方式。其中,有形或无形的科学技术创新元素逐渐升华融入一个新的世界,在此,建筑、艺术和设计相互交融影响,为人类设计满足需求且具备娱乐及其他多样用途的人造物。

妹岛和世(Kazuyo Sejima)在展览中呈现了一系列展望未来的实验性作品,吸引着我们,让我们融入其中。这些作品并不进行自我赞美,而是表达了对未知的好奇,也许这些作品就是我们想要的。

在此次双年展所选的主题中,涉及环境但并不像过去的"激进环保主义者"所表达的一样,所以环境主题并不明显。对妹岛和世(Kazuyo Sejima)而言,环境主题应当在科学研究之中,在成果应用之中,在可持续发展之中找到解决方案。她采用以人的需求为中心的方法,与那些敌视新技术的人的意识形态立场形成对比。我认为后者的观点是"愤世嫉俗"的,因为它违背了人类的观点,否认了创新可能带来的新的机遇。

妹岛和世(Kazuyo Sejima)再一次提出以人为本,而不是以产品为首。她理解并欣赏孩子们,因为在他们的世界当中,他们的感官和情绪并没有受到其它因素的影响,一直保持着自我的纯真。

交互设计 > 手势界面为用户提供浏览视频、音频的平台,用户通过手势指令在屏幕上操作,下载图片、文章、电影和音乐。又或借助蓝牙技术与数码相机的结合,利用超声波和红外传感器来实现对身体语言的解读,进而实现远距离的操作控制。

在电影《少数派报告》(Minority Report)中,汤姆·克鲁斯(Tom Cruise)望着屏幕,浏览着图片和地图,他用手滑动虚拟网络页面,这一场景似乎代表了一个遥远的世界。然而现在的我们已经实现了,甚至说是在过去就已经实现了。

同样的技术我们也运用于购物中心。当我们进行采购时,它刺激着我们的选择,为我们提供多种解决方案;当我们在办公场所、火车站、酒店前台时,它为我们提供大量的信息。它改变着我们的日常行为,改变着我们的社会关系。

这些应用程序属于交互设计学科,这是一个学习交互系统认知本质的设计领域,其基础理论制定于上世

混合现实

纪八十年代。事实上，交互设计的定义是由比尔·摩格理吉（Bill Moggridge，IDEO 联合创始人）和比尔·维普兰克（Bill Verplank，交互设计师及人因工程师）在八十年代共同创立的。他们是首批使用交互设计（Interaction Design）这一概念进行表达的人，超越了之前所谓的"用户界面设计"（User-interface design），从而创造了用户界面设计的新纪元，使得设计成为一门整合产品创意及开发生产全过程的学科。

之后的九十年代初期，交互设计已经成为设计实验和操作的特定领域。大卫·凯利（David Kelley），迈克·纳塔尔（Mike Nuttel）和比尔·摩格理吉（Bill Moggridge）于1991年创立了IDEO工作室，这个"实验室"提出以人为本的设计理念，并很快成为产品和服务体系设计的国际知名企业。

这是一种能够提供人类认知系统与信息处理工具之间交互应用的方法，在办公、教育、娱乐领域设计合适的具有多用途的产品。因此，他们的研究方向不仅涉及改善产品的技术功能，也包括满足用户的所有需求进而完善人们的生活。由此，用户友好型技术的理念开始逐渐兴起。

混合过程 > 混合体，正如我们所知，即两个或多个具有不同特征的物品结合杂交生成单一个体。在生物学方面，杂交是指通过两种动植物的混合，改变其中的部分特性来实现一个新的品种。因此，混合杂交就是一种形式向另一种形式转化的过程。奥维德（Ovidio）在他的《变形记》中以一种更富诗意的方式阐述了这种变化的过程，这一过程除了想象，还涉及了homo faber的行为，借助技术实践生成新的物品。变化这一主题，无论好坏，始终处于人类进化的过程当中，失去变化，人类将无法继续发展。而设计则如同一门连接技术与文化的学科，试图控制着其内部自发却又无可避免的混合过程。过去，设计在不同的时期与地区，利用特定的材料与生产技术来阐述人类的需求，不断地生产产品来改善人类的生存条件。而如今，设计不仅要面对创新技术的快速发展，同时还受到多种因素的影响制约。比如，电子信息的发展、网络的使用，使得不同区域的文化具有同化趋势；全球化势不可挡，广泛传播的混合文化使得新产品被源源不断地生产出来，而此时时间与空间的概念也逐渐模糊；同时，无论是在衣食住行等基本需求方面，还是在家庭、工作、学习、休闲等方面，我们的行为与愿景都发生了深刻的改变。

总而言之，这些产品成为了我们物质文化的基本构成元素，与此同时也引发了多种混合的概念。

类型混合，指的是一种混合体系，它影响了产品本身的特性。基于产品蕴含多年的原有特征引入新的性能，可使其成为一种新的产品。此外，它也包括由两种或多种不同的产品类型混合而产生的一种全新的类型。

功能混合，指的是同一产品或同一类型产品的不同使用形式。这是一种发展中社会与先进工业社会之间的产品类型的转移。这些产品虽然保留了原有的构成特征，但在不同背景中依旧会产生新的不同的使用形式。

技术混合，指的是涵盖满足新需求的产品的混合体系。从新的需求出发，衍生出新的产品，借此来改善人类的生活，同时通过应用新的科学技术，让设计生产新的产品性能成为可能，在强大的设计领域中，在全新的产品之间使用技术转移来形成一种混合形式。比如，借助纳米技术，将高性能器械、交互元件植入微小型产品中。

尺度混合，指的是在设计以及生产过程中，产品的尺度标准所产生的变化。这包括了由小产品到大物件的特征形式的转移，它使得家用产品似乎是一件小型的建筑，而建筑又好似大件的日常用品。尺度混合还体现在电子技术在产品生产过程中的运用。借助数字化管理的器械，即使是极小尺寸的产品，其生产过程中的可控错误已不复存在。

语言混合，指的是设计产品不同的表现形式与风格。也许这是最值得探究的方面，因为设计产品具备的可沟通内容是构成我们物质文化的基本元素，同时，设计产品也具备了多元化艺术领域中的美学特征。

女神宣告：
"远离神殿，你们将遮住头部，褪去服饰，将伟大母亲的骨头抛于肩后。"
很长一段时间，人们并不理解其中的意思。
直到普罗米特奥之子安慰妻子道：
"也许我理解错了，又或者我理解对了。但无论如何，对女神都不是一种亵渎。
伟大母亲是大地；那么我们应当向后抛的骨头就是石头。"
人们服从着，一边走着，一边将石头抛于肩后。
而这些石头（谁又能相信它们见证着古老的时光呢？）慢慢地不再坚固，一点点地软化，逐渐出现了形状。
拉长它们，观察它们，隐约地看到了人的形状。
同时又好像能看到大理石，甚至是雕塑的轮廓。
不久之后，
被抛出的石子都变成了扔它们的人的模样。"
摘自奥维德（Ovidius）（公元前 43 年—公元 17 年）《变形记》第一册

电影设计 > 设计逐渐涉足极具戏剧性的电影领域，衍生出一种设计的表现形式。正因如此，我们开始思考电影与设计之间的关系。

电影是一种构建叙述的手法，也是一种集特殊效果、服饰、多媒体、视觉设计于一身的复杂的艺术。也正是因为拥有着伟大的传统、匠人的能力，以及众多如达尼洛·多纳蒂（Danilo Donati）、丹特·费雷蒂（Dante Ferretti）、卡罗·兰巴尔迪（Carlo Rambaldi）等奥斯卡获奖者，意大利电影业闻名全球。

混合现实

在国际范围中,借助斯坦利·库布里克(Stanley Kubrick)的伟大作品,就能够很好地证明电影的创新不仅体现于语言风格方面,还应当考虑电影的摄制技术。正如他的《2001:太空漫游》,虽然这是四十多年以前对于未来的畅想,却描绘出了至今看来依旧给人以未来感的场景。谈到设计与电影,随着如今的电脑及多媒体器材的普及,设计师和导演(如,索尔·巴斯 Saul Bass,蒂姆·波顿 Tim Barton)为观众打造一种全新的视觉体验,在试图传递各种情感的同时,将故事与产品融入电影作品中。在这样的背景下,英国导演彼得·格林纳威(Peter Greenaway)在米兰三年博物馆中进行了《号外!2000年意大利创造的前奏》这一展览,利用装置向观众呈现了一个多面墙体的投影空间。他在墙面上投射放映了意大利著名的电影片段,并在前方展示了似乎能够融入投影背景的产品。伊塔洛·罗塔(Italo Rota)设计展览,安德列亚·布兰茨(Andrea Branzi)进行策划,他们选择了一系列产品来表达对意大利设计的"念念不忘"之情:安东尼奥·卡布安诺(Antonio Capuano)的《天国之光》(La luce del cielo),帕皮·科西嘉托(Pappi Corsicato)的《超舒适》(Il supercomfort),大卫·法兰里奥(Davide Ferrario)的《生机勃勃》(La dinamicità),丹尼埃莱·卢凯蒂(Daniele Lucchetti)的《堆砌的民主》(La democrazia impilabile),马里奥·玛通(Mario Martone)的《活跃的剧场》(Il teatro animista),艾曼诺·奥尔米(Ermanno Olmi)的《伟大的简约》(I grandi semplici),以及西尔维奥·索尔迪尼(Silvio Soldini)的《伟大的资产和神圣的奢华》(I grandi borghesi e la sacralità del lusso)。

社会本质 > 这是一种意识的转移。无论是个人还是公共用品,无论是在南方还是北方,这种同化趋势在我们日常生活中随处可见。在所谓的全球村中,这种特殊的情况试图抹去传统的独特特征以及不同区域背景的物质文化,进而促进相似的日常用品的普遍存在。这一问题涉及社会和经济,政治与伦理,美学与文化,同时影响了创新技术和生产系统的发展。正因如此,这是一个复杂的问题,我们需要从特定的设计领域进行多方面的研究探索。在功能性、技术特征、形式内容方面,深入地研究我们周围的有形或无形的产品及其体系。

转移 > 毫无疑问,如今的人们开始具备环境保护意识,并通过介入工业产品生命周期来节约能源。比如在产品生产阶段,通过精细操作来节省材料和能源;在产品组装和拆卸阶段,简化组装和拆卸的规章准则;利用废弃产品的材料和可回收元件进行二次生产来完成新产品的组装。
可以确定的是,过去提及可持续发展产品,一般来说只是涉及环境因素,而如今通过发布新的规则方针,促进可持续生产,已然上升至国家战略。比如加快汽车的废弃,来达到推广催化消声器的目的。这一战略也进一步影响了企业在研究、创新和实验方面的投资。
早在1993年,菲亚特已经开始执行一个名为F.A.RE(Fiat Auto Recycling)的项目。事实上,它是针对废弃汽车的某些部件进行的"梯级回收",利用粉碎技术将聚丙烯的保险杠转化为原材料,用于生产车用风口管道（一级回收）和脚踏板（二级回收）；回收的玻璃用于汽车以外的行业，大多用于生产玻璃器皿；座垫填塞物则被回收生产为建筑装饰地毯。在"梯级"的最后一个阶段，将反复使用过的无法再次回收的材料，作为工业高炉燃料使用。

微型设计 > 纳米技术为复杂且肉眼不可见的产品开辟了全新的研究领域。正如能够修复组织损伤、清除病毒的微型机器人；或是能够阻断甚至逆转老化的微型有机体。在生物学领域，纳米技术与基因操作技术相结合，带来了全新的变革；在材料科学领域，同样有着令人着迷的前景。纳米技术与化学、物理、电子工程相结合，发挥它们之间相互作用的最大可能性，以此生成新的具有电学、光学、热学和机械学性能的材料和器件，并能够提前设定它们的性能参数。举例说明，有机发光二极管采用特殊装置，在两片导电层之间加入一层有机膜，将电信号转化为光信号，来达到照明领域高度的设计自由。另外，由玻璃或塑料模具支撑而成有机聚合物膜，具有特殊的光学属性，可以此为材料制造智能玻璃用于汽车工业、建筑业和农业，并大大降低了生产成本。

纳米技术可以赋予材料以数字化形式，引入材料科学，这种发展与引入数字化类似。少即是多（Less is more）的原则，与起初相比，如今呈现出了更多的含义。少量使用材料，减少能源消耗，它更加注重产品的环保概念，试图在自然与人工当中寻找新的平衡；在加工工艺的微小误差方面，在去除冗余形象和无趣形式方面，它变得更有控制力；它促进了精细技术产品、小型产品和无形产品的设计与生产，利用微处理器技术让产品功能更具智慧性，仅仅通过简单的人体工学手势，不需要很高的文化程度也可以使用这些产品，使得人机交互更加便捷；它最终也成为一种研究方向，被运用于在未来场景中，精简物品本体。

§

[2]

p.79

文化创新

色彩社交
摘自《色彩描述设计所以》，2012 年《DIID》第 53 期
语义价值
摘自《在相通与差异间的设计》，2009 年《DIID》第 42—43 期
设计院校
摘自《健康设计》，2007 年《DIID》第 27 期
结合
摘自《罗马的拉齐奥地区：设计体系》，2006 年《DIID》第 21—22 期
生产领域
摘自《意大利设计研讨会》，2006 年《DIID》第 26 期
意大利风格
摘自《传播意大利设计》，2011 年《DIID》第 49 期
意大利印象
摘自《意大利制造的本质》，2004 年《DIID》第 10—11 期

意大利设计是一份错综复杂的宝贵财富，这离不开意大利设计的复杂性与其领土的多样性。然而，意大利设计也是一种广泛而坚实的社会现象。它通过共同的"文化创新"这一智慧方案，将与过去相关的恒久不变的特质保留至今。这一方案让不同代人之间的想法得以连续，即让大师、艺术家、年轻设计师、无名设计师和企业的作品相互联系，来展示历史是如何将最原始的物质文化传承给我们的。

色彩社交 > 二十世纪中期，彩色透明塑料兴起。慢慢地，纯白色半透明塑料得到大家的广泛关注，随之而来的是由此产生的一种全新的技术与色彩之间的基本联系。最近，苹果公司将色彩技术也纳入专利申请，这并不是偶然。

众所周知，上世纪70年代，艺术的跨学科，设计经验的渗透，语言表达的叠加，从中诞生出一门独立自主的设计学科。

在风格之间，形式与种类之间，工业与手工之间，甚至是实用艺术与概念艺术之间，色彩，都是一个无法忽视的根本因素。随着70年代激进设计的普及，色彩甚至成为了社会与文化的宣言，阐述了期间产品设计的实验运动风潮。后现代主义引入随意结合的艺术语言，反映了以大众消费和大众沟通为基础的波普文化。

以阿基米亚（Alchimia）和孟菲斯（Memphis）为例，虽然这个例子被过度使用，但是它是最能体现问题的。他们对绘画色彩和多彩、合成及自然纤维的使用，时而相似时而不同。他们的作品搭配上色彩，逐渐成为一种具有表演性质和叙述能力的物质。而此时，在安德列亚·布兰茨（Andrea Branzi），马西莫·摩洛希（Massimo Morozzi），克尼奥·德里尼·卡斯得力（Clinio Trini Castelli）的帮助下，随着《色彩发动机》（Colordinamo，1975-1977）一书的出版，色彩逐渐成为一种物质，它开始定义一种全新的设计学科。

在设计中被持续不断地使用的色彩，开始赋予设计以重要价值，其中包括所谓的产品的"柔软性能"。即使兼收并蓄、暧昧模糊，意大利设计依然具有"绘画"维度，在某些情况下，甚至在工业产品形而上学方面有着强烈的主导性。此外它还将产品视为一幅画作，引入精简的"二维"思想理念。无论如何，这只是概念上的转变，想象一下纸张之上浪漫的"红色"，就好像是绽开的罂粟花一样动人。

语义价值 > 意大利设计学院具有研究现象学并总结学术理论的特性。意大利的工业化直到上世纪五十年代才全面完成，然而就在那些年间，所谓经济繁荣碰撞上广泛分布的建筑与艺术的世界遗产，使得意大利设计在全球范围内得到了广泛的肯定。

具有家族特征的微小型企业广泛分布于意大利不同的区域，形成了一个共同体系。在这个体系中，大量的手工技能相互促进相互完善。大量的经济学家研究"意大利制造"这一普遍现象，并称之为工业区系统，这就是典型的意大利设计风格。

另一典型风格则是意大利设计文化的普及。无论是技术学校还是美术院校，无论是建筑学院还是理科学院，最重要的是通过每天不断地接触来感受我们周边存在的大量的"美"，赋予大众的"高品味"也成为意大利人宝贵的财产之一。

当然，创造力也与世代相传的行业具备根深蒂固的良好实践紧密相联。这些行业大多始于罗马时期、中世纪时期的大型工厂，直到从布鲁内莱斯基（Filippo Brunelleschi）到布拉曼特（Donato Bramante）的

文化创新

文艺复兴时期,从贝尔尼尼(Gian Lorenzo Bernini)到博罗米尼(Francesco Borromin)的巴洛克时期,诞生了一系列的工种:灰泥工、模型工、车工、裁缝师、裱糊工、大理石工、壁画师、石匠、雕塑师、画师、装饰者、铁匠、木匠和无与伦比的艺术家,他们分布在各个"公司"中,除了教皇和红衣主教以外,他们还服务于皇帝、王室以及贵族。

同时,在贵族的资助下,诞生了不少宏伟壮丽的艺术作品,最重要的是,也滋养了优秀的艺术家的艺术才能和卓越的创造技能。随着工业化的发展,意大利诞生了许多新兴行业。丰富的历史文化遗产融入意大利物质文化社会,随之而来的是经济繁荣时期,在这样的背景下诞生了设计行业,大量的魅力四射的产品也随之出现。

产品具备的功能性、其语义价值、其传递的情感,以及由柔和多彩的曲线、欧几里德定律线条和几何图案引发的熟悉度,都一度受到大众的追捧。这些由复杂的制造系统生产,或经过简单装配而成的产品成为大众的欲望产品,正如当时的菲亚特500汽车被视为是最性感的车型一般。

在意大利设计中,想象力总是伴随着严谨的设计,并将功能与戏虐,人类情感与复杂的平衡相结合。所谓复杂平衡,是指产品美学与谨慎对待新型材料、恰当技术,以及所有面向未来的创新实验的研究之间的平衡。在意大利学院有诸多大师,每一个都是独一无二的。在此,我认为介绍布鲁诺·莫纳(Bruno Munari)是十分恰当的。他借助 homo faber 的精神来丰富其视觉艺术和设计经历,他具有大量的知识储备,涉及艺术、应用科学、哲学、教育学、文学,他还经常游历东方世界。因此,可以说布鲁诺·莫纳(Bruno Munari)是一个非常饱满且极其复杂的人物形象。他还总是试图为人类和儿童世界提出各种创意想法。如今的意大利设计学院,构成了统一的国家大学教育系统,并一直处于欧洲教育的最前沿。借助持续的方法论和文化的主线,来帮助我们适应它们的转变,尤其是通过不断地推陈出新、主要的经济转型和大型文化运动,在传统与创新之间形成一种平衡。

设计院校 > 提供众多的设计课程,是为了响应大量年轻人想要从事设计行业的需求,更是为了满足企业对专业人才的渴望。而今,那些传统的院校已经无法储备这样的混合型人才。事实上,这些设计院校(或许是唯一的)将新型技术引入教学机构,引发了深刻的变革。新型技术改变了学习方式,使得获取知识的途径更加多样化;改变了设计技术,使得模拟更加完美;改变了设计与生产之间的关系,使得远距离遥控数控机床进行精确操作成为可能;在生产过程中引入了机器人来替代传统的制造机器;引入能够改变人体性能,甚至外形的交互产品,进而改变了用户与产品之间的关系。

设计院校的教学任务不仅涉及了理论知识,还包括实践活动,新的学科领域决定了理论与实践思维的界限。事实上,设计院校的教学组织如同跨学科领域一般。

鉴于有形或无形的产品的大量涌入,设计者需要掌握更新的知识,具备更综合的能力以及技术操作技能。而设计院校显然知晓培养设计师的真正需求,总而言之,它就是这样一种教学机构。

在设计院校中，所有的教学课程围绕着"信息"这一核心问题展开。这一核心问题表现了创新信息技术的潜力，它改变了我们思考、认知及表达沟通的方式，它在产品的生产过程，交互使用及最终消费方面也产生了巨大的影响。

结合 > 拉齐奥大区，乃至罗马地区的生产体系在"意大利制造"中扮演重要角色，这是毋庸置疑的。

之所以重要，是因当地很好结合了意大利风格的三大元素：传统性、创造力和卓越感。同时，也是因为当地广泛分布的生产活动已经形成了完善的体系，在这里充满了机遇与地方特色，在这里艺术与创新技术融合，在这里历史底蕴衍生出新的行业，新的致力于传播文化和旅游的企业。也是因为当地的品牌产品或才华设计师设计的著名作品，比如从Valentino到Capucci，从sorelle Fontana到Biagiotti的服装品牌，以及像Galante和GrimaldiGiardina这样的新兴品牌。也是因为当地那些满足日常生活需求的无名产品，正如奇维塔卡斯泰拉纳工业区的产品，以及从波梅齐亚发展到拉蒂纳地区的复杂的工业体系产品。也是因为当地的应对全球挑战的高科技含量产品，其中涉及电信行业和太空领域。也是因为当地的化学制药和生物技术领域所表现出的巨大潜力。

除了中小型企业结构以外，文化体系和教育体系也是增加当地竞争力的重要元素之一。科技中心、研发中心、实验中心形成了完整的体系，在创新方面给予产品生产系统以最大的支持。之所以重要，还因为当地自身的特殊性。除了生产制造体系以外，还拥有完整的从事服务设计的企业系统，以及众多生产无形产品的公司，它们涉及许多新的专业技能，比如视听领域或通信工程等。

通过加强整个地区的考古资源与文化遗产来促进经济的发展。例如在罗马音乐大礼堂、圣切契里亚音乐学院开展相关活动；与罗马和其它地区的博物馆系统共同组织研究。围绕这些活动的同时，大中小型企业相继发展，逐步涉及视觉传达设计、平面设计、多媒体设计、展示设计、装饰设计、公共空间设计以及舞台特效设计。为了提升影视化领域的专业性，他们不断地与RAI广播电视台、Cinecittà影视基地以及罗马歌剧院进行长期合作。无论是舞台布景、服装造型，还是灯光设计和舞台特效，众多的奥斯卡奖项可以证明，这些历史足以享誉全球。影视化的根源甚至可以追踪到罗马歌剧院的储藏室，除了舞台布景以外，保存的近五万套的舞台服装成为了二十世纪意大利艺术历史的重要见证。此外，在商品的设计与生产领域诞生了许多新型企业。比如Enzimi，Zètema以及Civita，他们联合国家级博物馆MACRO和MAXXI促进传播并组织活动来推广文化遗产，策划艺术展览。

在设计领域，教育体系也受到越来越多的重视。例如，罗马大学提供了从本科到研究生，再到博士的众多课程，全面涵盖了新兴设计领域多元化的专业需求。设计课程针对日益复杂的生产体系，开设了广泛的对应专题。事实上，工业产品设计师逐渐成为一个非常复杂的职业，因为他们不再是简单地考虑产品的外观美学，还需要掌握更加复杂且专业的主题观点，从有形的设计，到基础设施，甚至到无形的服务

文化创新

领域。提到视觉传达设计、多媒体设计以及交互设计，就是使用适合的语言、工具以及必要的技术支持，有可能还会涉及平面印刷、平面包装、品牌形象、动画图像，以及网络中使用的界面图标。随着微电子技术的引入，信息技术、远程信息处理技术、机器人技术、纳米技术和新型材料的普及，工业产品设计逐渐丰富了产品的性能和通讯内容。因此，这也需要设计师掌握多种技能，来适应不断更新发展的设计生产模式。实际上，设计必须根据生产策略以及市场需求来进行管理和规划，这也是设计管理和设计指导的职责所在。显而易见的是，这些新兴领域与研究中心紧密合作形成了一个真正的区域研究空间：不仅有太空机构的战略项目，还有伽利略试验场的活动，与此同时，不仅研究对激光技术的使用，还研究新型材料，研究纳米技术及其应用。

正因如此，无论是有形的产品还是无形的项目，拉齐奥大区的设计文化都展现出了自身的卓越水平，创意、创新与传统的结合为其加分不少。

生产领域 > 意大利设计已然成为一个多元化设计体系，在全国范围内形成了清晰且复杂的系统。因此，现今的意大利设计不是单一的，而是多方面多样化的。它呈现出多元化特性，被广泛运用于不同的专业领域，涉及不同的生产区间。

意大利设计涉足汽车领域。在马拉内洛的超凡工作室设计出了具有超强审美、超高技术含量的法拉利汽车，而在不远的地方就是著名的博洛尼亚汽车设计地区。

意大利设计涉足船只领域。在里雅斯特，Fincantieri 造船厂为世界各地的客户定制大型可以横渡大西洋的船只，其它船只制造厂在全国范围广泛分布（如，位于 Ancona、Forlì、Torre Annunziata、Fano、Cattolica、La Spezia、Sarnico 的 Gruppo Ferretti；位于 Fiumicino、Sabaudia、Gaeta、Posillipo 的 Gruppo Rizzardi；位于 Messina 的 Gruppo Aicon）。

意大利设计涉足家电领域。家电是 Indesit 集团研发中心（意大利设计中重要的一部分）的重要部分，在法布里亚诺和耶西所处的机械区域，至今依然存在着像 Elica Group 一样的杰出企业。

意大利设计涉足制造领域。在意大利东北区域分布着众多制造业，其中包括的机械和分包业务，中小型企业在生产纺织机床或精密机械方面处于全球领先水平。

意大利设计涉足创新技术。在威尼托大区附近有着众多先进技术研发实验室，其中包括纳米技术的研究及应用，威尼托纳米科技实验室就将其应用于微小型产品中。

意大利设计涉足航天航空领域。除了皮埃蒙和坎帕尼亚的跨区合作之外，在拉齐奥大区 ASI 和 Alenia 周边形成了新兴的航天航空研究区域。

意大利设计涉足影视化领域。罗马在视听领域具备的悠久卓越的历史，为日益先进的研究奠定了基础。

意大利设计涉足电影领域。在装饰、舞台、服饰方面，罗马的悠久历史和国际知名度相得益彰，这点，从无数的奥斯卡奖项就可以看出。

[2] 设计 _ 文本与语境

意大利设计涉足时尚领域。不仅是罗马,时尚所波及的范围还涵盖了佛罗伦萨、米兰。除了历史悠久闻名于世的 Valentino、Capucci、Armani、Versace、Ferragamo、Biagiotti、Prada、Missoni,还诞生了众多年轻的设计工作室,并逐渐形成生机勃勃且生效显著的产业链,给全球带来巨大的影响。
此外,意大利设计还涉足其它典型的"意大利制造"领域。

意大利风格 > 传播意大利设计是一项艰巨的任务,若你要在杂志的几页进行书写,那么就是难上加难了。但是定期进行归纳总结,即便不够详尽,对我也是有益的。为此,我们设定了一个具体的时间区间来进行阐述,就是两千年的首个十年,即2000年至2010年间。
这十年来,意大利设计逐渐成为一个多元化设计体系,表达了一系列的实际生产与文化,即使处在全球化日益剧烈的背景中,依旧保持了自身的独特特征。正如意大利设计教育领域,广泛分布的设计学院形成了遍布全国的完整体系。
这十年来,诞生了许多设计新星,无论他们是意大利人还是外国人,他们始终与意大利生产体系保持着紧密的联系。在某些情况下,他们追随着老一辈将意大利风格发扬全球的设计大师,学习着他们的设计方法和设计文化,在一定程度上实现了设计理念的延续。但更多的时候,设计学科上出现了一种断裂,一种不连续的状态。全球化的到来,消除了上世纪难以消弭的国家文化与传统之间的距离,因此,一种全新的设计方式应运而生。当今艺术、设计与建筑之间的界限越来越模糊,它们相互作用、相互影响、相互混合,形成了一种全新的有形或无形的表达方式,因此,我们应当以一种全新的视角去看待现在与未来。
这十年来,设计跨越了传统的领域(家具设计、室内设计、产品和交通工具设计),在其他领域风生水起,如展览设计、公共空间设计、时装设计、视觉传达设计及多媒体设计等等。
这十年来,媒体对设计学科的兴趣也日渐增长,只要想想多少报刊杂志为设计开辟了专栏主题即可。
这十年来,为传播设计文化设立了众多机构:从 Musei d'Impresa 到 Collezione Farnesina,从 Fondazione Valore Italia 到米兰国际家具展(Salone Internazionale del Mobile),从 Abitare il Tempo 到米兰三年展(Triennale di Milano)的意大利设计博物馆开幕式。"什么是意大利设计?"米兰三年展的前三个展览巧妙地回答了这一问题:《意大利设计的7个念念不忘》(Le sette ossessioni del Design italiano),《批量与定制》(Serie fuori serie)以及《我们是什么?》(Quali cose siamo)。这些日子,米兰三年展正在进行它的第四个展览《梦想工厂》(Le fabbriche dei sogni),展示了意大利设计工厂中的人类、思想、企业与矛盾。
尽管人们越来越重视设计在传统领域的应用,但对高性能产品所表达的设计却一无所知。那些创新不仅涉及外形,还丰富了产品的种类。不仅涉及满足人类需求的多样化的产品,还包括那些日常的无名产品设计。

文化创新

人们更多地关注设计领域的领军人物，虽然他们是当之无愧的，却忽略了新兴设计师。他们的新型设计往往更多地关注家庭以外的空间，如人类与家居之间的狭缝，人类需求与感官情感之间的共通，或直接涉及人类与物品之间的关系，使其成为新的伦理价值和新的愿景的一部分。

我们仍然记得并时常提醒自己，意大利设计时代是历史的结果，但并不是那种常见的有点严肃而又自我独立的历史。而是像朱利奥·卡洛·阿尔甘（Giulio Carlo Argan）在1982年写的精彩文章中描述的一样。这篇文章是对《意大利的再次演变——八十年代意大利设计》展览（Italian Re-Evolution. Design in Italian Society in the Eighties）的目录式介绍，在文章中详尽地描述了意大利设计风格的特征及其根源。

在这首个十年中，我们并没有忘记阐述意大利设计的两大特征：制作的智慧和产品的标志价值，它们能够在产品设计中将传统手工文化与创新文化相结合；我们并没有忘记阐述在意大利风格中是如何运用制作的智慧将创造力转化为设计方案的，也形成了意大利设计历史的所有产品的标志性、特殊性力量，这在过去的十年中依旧是值得一提的附加价值。

意大利印象 > 1982年，朱利奥·卡洛·阿尔甘（Giulio Carlo Argan）为《意大利的再次演变——八十年代意大利设计》展览（Italian Re-Evolution. Design in Italian Society in the Eighties）书写了一篇名为《意大利人的设计》（Il design degli italiani）的目录式介绍。从标题就可以清楚地了解这篇文章的主题和背景。即使工业设计非常复杂多样化，我们依然可以对特定的生产与文化领域进行分析。为了深入地理解"意大利制造"这一现象，从这种方法着手似乎是极其恰当的，因为它基于"意大利制造"有着自己文化特征的批判意识，这对于分析这一现象而言是十分重要的。这也是我们为什么要借助工业产品的技术、美学、经济方面的因素来理解"意大利制造"，这是有益的，这也是和朱利奥·卡洛·阿尔甘（Giulio Carlo Argan）在文章不谋而合的观点。

第二次世界大战结束后，意大利摆脱了文化孤立，重新启动了一个与抽象社会相对立的工业生产模式，付诸于现实生产当中。在此期间，诞生了许多小型企业，它们都有一个共同的特点：对日常物品的种类进行再创造，正如朱利奥·卡洛·阿尔甘（Giulio Carlo Argan）所言，即寻找产品的标志性特征，以及寻求它们刺激触觉、视觉等感官的能力。

从这个方法开始，意大利印象（Italian Look）很快流行起来。在意大利印象的演变当中，其发展往往取决于产品的可识别性。即使在产品中运用高科技来丰富其性能，但是最重要的特征依旧是其外观。所有的产品都试图被看见。即使在厨房和卫生间，传统上因为过于注重技术和性能而被忽视外观的设备，也开始变得美观起来。渐渐地，新的材料改变了周围环境、产品及其形式。质地轻便柔软、色彩透明鲜艳的塑料，慢慢地改变了产品外观及其在家庭空间中摆放的位置，同时，也改变了大众消费文化。

随着风格语言碎片化，社会经济危机的产生，意大利设计危机开始初见端倪。这是伴随意大利衰退的一场危机，此时的意大利在不同文化科学领域，在艺术与技术，经济与政治之间已然无法进行良好操作。人们拒绝正视衰退这一现实，虽然提出庆祝、保护、重启"意大利制造"，但是工业生产体系也正处于一个显而易见的危机当中。

意大利制造总是与农产品及食品行业息息相关，这也进一步代表了这一古老传统的国家，描绘了意大利文化或其产品（比如法拉利）。事实上，意大利制造的产品具备了综合各个国家不同文化所带来的多样化能力，而在意大利只是进行了生产或组装。那么，意大利制造对我们而言究竟意味着什么？是可以运用我们自身的能力、文化和创造力来进行生产制造？还是仅仅是一个让产品具有附加价值的品牌？我们应当将意大利设计根植于当地传统之中还是将其融入复杂的创新信息的全球化网络之中？意大利制造的**概念**是否能够和思想流通的全球化趋势兼容？又或是和所谓的本土化生产模式共存？这也值得我们仔细思考。

§
[2]
p.87

传统与变革

一堂课
摘自《设计无处不在》，2013 年 1 月讲座
建筑元件
摘自《可期的智能》，2014 年《DIID》第 58 期
新的刺激
摘自《在相同与差异间的设计》，2009 年《DIID》第 42—43 期
地中海——两地之间
摘自《汽艇》，2009 年《DIID》第 40 期
大众设计
摘自《健康设计》，2007 年《DIID》第 27 期
美化
摘自《产品的装饰与美观》，2006 年《DIID》第 23 期
无名
摘自《众多或无名产品的力量》，2005 年《DIID》第 15 期
产品流
摘自《污染、传递与同质》，2005 年《DIID》第 14 期
复杂的文化
摘自《相同与不同，共同市场设计》，2005 年《DIID》第 12 期

人与人之间，一代又一代，不断地回忆见证着社会事件、仪式集会、文化传承，这些汇集成设计传统。然而，传统通过文化沉淀巩固自身，随着知识的改变，传统也在发生变化，这些变化更新是持续且偶然的。事实上，这与设计方式的改变如出一辙：它正处于过去与当下的沟壑之中，它面临的这些改变是自发的，或是所处内外因素引发的。因此，设计是将过去的回忆，当下的状态以及未来的渴望相平衡的一种模式。

§
[2] 设计 _ 文本与语境

一堂课 > 我一直在找寻优秀的书籍来充实我的写作内容，让这本书能够更加浅显易读。机缘巧合，我读到了这样的一段文字。这是一位俄罗斯哲学家、数学家在1917年写的一篇名为《课堂与学术课堂》(Lezione e lectio) 的文章。他说："如果你把两个思想相近的人之间的一段对话定义为一堂课，也不算太错。一堂课并不是一段行程，你坐上电车，选择最近的路程，随着轨道它带着你走向固定的终点。而应该是一段徒步旅程，虽然也有着最终目的地，但是在旅途过程中终点不是唯一的目的。选择哪条路并不重要，重要的是最终到达了。因此，我们可以缓步前行，欣赏着沿途风景，哪怕是一块石头、一棵树、一只蝴蝶，我们也可以停下脚步静静观赏。如果想起什么，我们甚至可以掉头回转去寻找遗失的美好。"也许他的这种说法有些过于夸张，但是这个旅程的言论却得到大家支持，并被广泛分享，也因此确立了教学的基本原则：一堂课，也就是两个思想相近的人的一段对话。

建筑元件 > 这些年来，我们的设计文化中不断充斥着"智能"的概念，它似乎已经成为用来支持可持续发展和综合现代化的补救术语，使得我们所处的城市网络遍布，形成一个实用且紧密相连的场所。尽管如此，这还远远不够。当我们想到在智能文化概念下设计生产出的住宅和产品，它们利用通信技术，却无法寻找到新的智能形式、物理构建来满足人们的新的审美，人们的各种需求，就能得知"智能"概念往往有些夸大其辞。总而言之，相对于真正重新审视过去的设计模式，相对于实验创立新的产品或服务的范式，相对于探究产品内部改变的真正原因，这些应用于项目前端的科学知识更像是一句简单的口号。

最近我参观了雷姆·库哈斯 (Rem Koolhaas) 策划的威尼斯建筑双年展，其展览主题为：建筑元素。在几期"歌颂当代"的展览之后，今年的展览以建筑基本元素为主题，展厅中布满了建筑元件的标准与原型，其中包括楼梯、窗户、屋顶等，而这些正是设计师们长期以来研究评论的对象。在展览大厅中，看着原型范式变革中材料与技术所发挥的作用，让我不得不思考这些新的技术应当怎样呈现。以"门"为例，我们都知道门的作用是保障绝对的安全，但如今加入的无形技术系统，比如金属探测门，就在一定程度上改变了门的物理形态。之后是"地板"主题，此时的地板不再是一块可以踩踏行走的平面，而是一个可以进行加热或制冷操作的结构系统。正如鹿特丹的"科技歌舞厅"所展示的，产生能量或借助磁场指导机器人行进（如，亚马逊仓库中的某些区域由机器人进行操作），好似标注了"人类禁止入内"的信息。

观察着展厅中布置的"壁炉"元素，我发现它们借助电子技术更加贴近人类的需求。一边是原始的形式，炉膛清晰可见，其中之一是西班牙瓦伦西亚当地的壁炉，展示了当地人民使用它的生活状态。另一边是未来化的高科技家用暖气设备：麻省理工感知城市实验室借助追踪技术，设计了被称为"本地工作"(local working) 的作品。它不再是需要加热整个空间来抵御严寒，而是一个"简单"的紧随着人移动的热量球。美国研究者的成果很好地阐述了应当如何靠近"信息技术"的意识形态，这是唯一真正在

传统与变革

空间服务概念中思考及使用技术的方式。
这一届以建筑元素为主题的双年展，既不是亵渎玷污过去，也不是不加批判地支持现在，而是提供对于未来发展可能性的指导。其中最为重要就是"电子信息"（作为智能的同义词）的引入，将其注入产品或系统，使得这些物品能够"聆听"、"诉说"和"传达"。尽管如此，在"理解"和"欣赏"这个赋予产品以生命的过程的同时，除了思考它们的实用性，我们还不得不反思它们的可取性。

新的刺激 > 至今我们依然沿用现代运动时期所定义的设计师的社会职责，不同的是，设计与自然，与特殊背景之间的关系发生了改变。所谓自然，人们依旧在大自然中找寻形态灵感，激发新的美学主题。而所谓的特殊背景，不仅成为材料使用的参考，还成为了刻画产品形式的重要启示来源。
斯堪的纳维亚学院至今为止依然具有极强的身份特征，这离不开它们的起源文化，也离不开与大师之间的紧密联系。阿尔瓦·阿尔托（Alvar Aalto）及阿恩·雅各布森（Arne Jacobsen）作为学派创始人，依然强调与传统之间，与功能性和有机性之间的关系。设计产品，应当考虑在一个人与环境辩证平衡的空间中，它与建筑之间的关系，人类应当成为装配、构成住宅空间的主宰者。
由于历史和文化背景的差异，各国工业化方式与时间的不同，地中海学院在葡萄牙、西班牙、法国、意大利和希腊的身份特征体系也有所不同。
毫无疑问，法国学院在世纪之交受到伟大的工业设计运动根深蒂固的影响。更重要的是，巴黎是艺术先驱们重要的集中地，经济体系也随着工业革命的到来发生了根本性的改变。法国设计学院奠定了自己的基础，扎根于现代运动理性主义和功能主义的文化理论当中，并表现出了对人类社会和技术创新的巨大兴趣。实际上，装饰艺术沙龙展出的新产品，以及让·普鲁韦（Jean Prouve）所在的国立工艺学院（Conservatoire National des Arts et Métiers）都能很好地证明这一点。
设计不仅影响了技术的发展，也促进了生产体系的飞跃，更加改善了人类生存的现代社会的状态。正如柯布西耶（Le Corbusier）在《新精神》（Esprit nouveax）和《住宅是居住的机器》（Machine à habiter）中提及的理论，时至今日依然是现代建筑的真正宣言。
在这基础之上建立了法国学校的基本特征。在上世纪，homo faber 受产品构建的启发，合理运用材料，利用风格来突出产品的性能与形式，将技术转化为审美特征。而许多像让·普鲁韦（Jean Prouve）一样的大师巧妙地将手工艺、企业与 homo faber 的精神结合。Rene Herbst、Robert Mallet-Stevens、Eileen Gray、Sonia Delaunay、Charlotte Perriand，又或是皮埃尔·查罗（Pierre Chareau）的重要作品，构成了法国学校的根源。其中，皮埃尔·查罗（Pierre Chareau）的唯一作品《玻璃之家》（la Maison de Verre），表现了工业产品、艺术作品和居住空间之间的紧密关系，如果没有现代性思想，那么它们无法结合并表现出崇高的工业文明以及超凡的现代特性。
西班牙和葡萄牙对设计的兴趣并没有在世纪之交爆发，因此伊比利亚学院仍然需要更多的刺激来促进学院制度的完善。但是我们可以说，在这个地区的设计受经济和后工业文化影响，几乎跳过工业期，即使经历过也没有真正面对这一时期。

新的刺激不仅来源于大众产品，还在全球化背景中受到当地文化的影响；不仅来源于对设计概念的解读（设计不只是一个与产品相关的学科，还是为探索科学服务的工具），还受到创造传播新事物的方式的敏锐眼光影响；它是当今时代和方法论变革中的伦理设计，成为为社会可持续发展和为大众消费指明方向的北极星。

新的刺激还与伟大的传统手工和应用艺术紧密相连，这些同样体现了葡萄牙城市的空间质量。在过去，传统手工及应用艺术以一种强烈的表现方式形成了城市环境和家庭空间的主要特征。比如，高质量可拼接的瓷砖画被镶嵌在当地的建筑外墙上或城市道路上，又或者黑白色的岩石片（玄武岩或花岗岩）出现在葡萄牙城市的街道上形成一幅马赛克作品。

此外，现代主义传统文化也在西班牙作品中体现得淋漓尽致。很显然，一个具有强烈唤醒特征的设计正在西班牙蓬勃发展。所有产品都在不断地刺激你的感官，让你试图去触碰感受。这些欢乐的产品激发着人类的情感，它们是创新、前卫思维的结晶，同时，它们也向我们描绘了地中海上的耀眼光芒。

地中海 —— 两地之间 > 一个地中海国家，就存在一种设计。我们想要了解设计是如何在其领域表现物质文化，通常会以欧洲作为中心视角，因为欧洲具有灿烂的历史和文化。但是相反，我却想要从欧洲以外的地中海地区去深入，也正是因为当地的历史和文化。

之所以将伊斯坦布尔作为中心，首先从地理位置出发，它是唯一一座横跨亚欧大陆的城市。此外，还因为它的政治和历史，从拜占庭帝国到奥斯曼帝国，它都是重要的国都。古往今来，伊斯坦布尔具有无与伦比的文化体验，渗透了整个奥斯曼帝国，它是一个真正伟大却不总是被认可的地方。大多人研究文艺复兴时期伟大的洛莱佐，但极少人探讨苏莱曼一世时期的璀璨文化和政治，以及建造了清真寺系统的建筑家米玛·希南（Mimar Sinan）。这些建筑物在伊斯坦布尔海峡和金角湾两侧连绵不绝，形成了完美的城市风景线。米玛·希南（Mimar Sinan）是有史以来最重要的建筑家之一，我们真的应该将他设置在我们建筑学院的课程中，来更多地学习他的作品。他与米开朗琪罗（Michelangelo Buonarroti）和帕拉迪奥（Andrea Palladio）处于同一时期，半个世纪以来，他作为建筑师和工程监督者，仿佛造物者一般，建立了奥斯曼帝国几乎所有重要的建筑物。

相隔一个世纪的两位作家，埃迪蒙托·德·亚米契斯（Edmondo De Amicis）和奥尔罕·帕慕克（Orhan Pamuk）都曾在伊斯坦布尔海峡上创造了自己的作品，描述了这个具有多元文化、多种族的历史古城。埃迪蒙托·德·亚米契斯（Edmondo De Amicis）描述加拉塔大桥，奥尔罕·帕慕克（Orhan Pamuk）回忆童年，一个观察人，一个观察物。埃迪蒙托·德·亚米契斯（Edmondo De Amicis）观察着加拉塔大桥上来来往往的人流，尽管他们都是奥斯曼帝国的臣民，但从他们的身体特征，所使用的语言以及他们的随身物品就很容易分辨他们来源于欧洲还是亚洲。就好像一支多肤色多种族的商队，混合了不同的民族、文化、宗教、习俗和传统，在桥上来来往往川流不息。

奥尔罕·帕慕克（Orhan Pamuk）在《从拜占庭到伊斯坦布尔——两大洲之间的港口》（De Byzance à Istanbul_Un port pour deuxcontinents）展览闭幕时，为目录册书写了一篇文章。海上飘荡着的一群老旧的汽艇，它们往来于伊斯坦布尔海峡长达半个世纪之久，奥尔罕·帕慕克（Orhan Pamuk）观察着其中一艘，就仿佛回到了自己的童年。他缓慢地靠近海岸线，开始与自己竞相追逐，试图在海峡边的城市中，在纵横交错的马路上，高低林立广告牌中，星星点点的窗户里，混沌模糊的建筑物中找到一个与众不同的特征，一个自己熟悉的符号，甚至是自己的家，这些都足以让一个孩子安心下来。这显然是以一种饱含诗意的方式来描述独特的多样性。这也正是当地的特色，即"两地之间的大海，地中海"（in mezzo a due terre）。当伊斯兰征服了整个非洲海岸和部分伊比利亚半岛时，这个概念在7世纪首次出现，因此"我们的海"（Mare Nostrum）这一老旧称谓则不复存在。

地中海，正如费尔南·布罗代尔（Fernand Braudel）描述的那样，"万千事物汇集于此，它不仅是一种文明，更是层层文明沉淀积累。游历于地中海，意味着在黎巴嫩遇上古罗马时代，在撒丁岛遇上史前史时期，在西西里遇上古希腊城市，在西班牙遇上阿拉伯文化，直至追溯到马尔他的巨石建筑、埃及的金字塔。所有的古老神秘交融于此，千百年来，驮畜、马车、货品、船只、思想、宗教，甚至是生活方式纷至沓来，让历史更加错综复杂。当然，其中也包括植物，我们认为是起源于地中海的植物。事实上，除了古老的橄榄、葡萄树和小麦，其它都是来自大海的另一边。那些深绿色灌木的金黄色的果实橙子、柠檬、柑橘是阿拉伯人从遥远的东方带来的，龙舌兰、芦荟、刺梨来自美国，那些有着希腊名字的桉树却来自澳大利亚，柏树来自波斯。尽管如此，这些植物终究成为形成地中海风景的重要元素。自然景观遇上人文景观，在我们的记忆中，不规则的地中海形成了一道连续的风景线，所有事物融合重组形成一个整体"。

所谓"另一个地中海"（altro mediterraneo），就是指从南边望向地中海的区域（北非地区），众多差异在此汇集同化为一个整体，我们试图从这里去了解设计（另一个设计）。"另一个设计"所表达的创意维度，其根源可以追溯至千年以前的传统文化，从中东地区跨越海洋直至非洲地区。

借助当地新旧传统交替混合，"另一个设计"的产品远销至欧洲地中海地区。通过千丝万缕的无形网络，融合了来自不同区域的风俗习惯，形成了自身独特的风格。

有很多例子可以说明，其中之一就是茴香酒。这是一个很有趣的例子，因为它是一种在地中海所有港口城市都很普及的饮品，它在不同的地区有着不同的名字：西西里称之为Tutone，法国称之为Pastis，土耳其称之为Raki，希腊称之为Ouzo，中东称之为Arak，意大利称之为Sambuca，其实它的名字来源并非偶然，当时茴香是由东方漂洋过海来到奇维塔韦基亚港口，因此是从阿拉伯语中的Zammut衍生而来。

另一个例子则是地中海地区陶瓷生产的传播和发展。古老的依兹尼克陶瓷，装饰了从拜占庭帝国到奥斯曼帝国众多雄伟的建筑。同样的生产方式、形式、装饰色彩,也可以在意大利维耶特里陶瓷和葡萄牙瓷砖画中找到。

同时,"另一个地中海文化"也是欧洲当代艺术的灵感来源,欧洲艺术家们重新审视另一个地中海的物质文化,从中选取灵感元素以一种新的形式在新的场景中进行表达。保罗·克利(Paul Klee)作为当代绘画杰出人物之一,也是包豪斯(Bauhuas)超凡教师之一,在这方面堪称典范。他多次前往埃及和突尼斯,这些旅程改变了他作品中的色彩、平衡和几何构图,虽然他借鉴了抽象的手法和表现形式,但是旅行过程中视觉体验的影响是显而易见的。

地中海东南部的沿海国家的设计,表达了一种介于传统和创新愿景之间的辩证关系。这种创新并不是表现在复杂的高科技、产品性能或制造系统的使用上,而是对改变产品的形式,使用新型材料,特别是在纺织品及其使用方面。事实上,其中不乏成功的时装设计师,比如具有土耳其塞浦路斯血统的设计师候塞因·卡拉扬(Hussein Chalayan),他是创新实验研究的设计奇才。比如在摩洛哥生活与创作的布料设计师軟米亚·雅拉勒(Soumiya Jalal)。比如东部的土耳其设计师阿齐兹·萨热耶尔(Aziz Sariyer),他创办了德林(Derin)公司。比如威达(Vitra)公司,他们的商业活动已经从地中海地区辐射至全球各地。然而,创新也代表了以新观念思考未来的愿望,为地中海周围国家提供了一种新的表现形式。即使这些国家的本土设计相互矛盾且有所不同,但都试图与全球化趋势进行正面交锋。

大众设计 > 设计文化的传播提高了我们周边产品的质量,从而改善了我们的生活条件。

欧洲的一些国家,例如北欧国家就是很好的例子。在这些国家中,设计是集体文化遗产的一部分,设计定义了我们周边产品的特征,设计甚至决定了公共空间、室内,乃至我们日常生活环境的普遍质量。

除了诸如阿恩·雅各布森(Arne Jacobsen)或阿尔瓦·阿尔托(Alvar Aalto)的大师以外,还有众多设计师对设计文化进行了诠释,他们满足了公众对空间质量的共同需求。无论是在公共场所、住宅、办公室,还是在学校、医疗区,都能感受到设计借助产品为日常生活带来的附加价值。

正因如此,家喻户晓的宜家诞生于北欧国家,这并不是偶然的。

美化 > 过去,"装饰"被视为产品的部分特征,并与其技术和功能息息相关。"装饰"大多用于产品的美观修饰又或是作为其形式和功能的附加物。从这个意义上,1908年,阿道夫·卢斯(Adolf Loos)在他题为《装饰与犯罪》的著作中提到,在人类物质文化的进化中,需要从使用的产品中去除装饰因素。但同样也存在与之相悖的著名理论:约翰·拉斯金(Jhon Ruskin)就倾向于将作品的艺术表现归功于装饰,他认为功能与技术特征只有使用价值;里格尔(Riegl)和弗洛永(Focillon)则认为,装饰表现了人类渴望按照既定的艺术原则来塑造自己所处的空间及所使用的日常物品,因此它与产品的形式和制造不可分割。

可以肯定的是，人类总是将其对各自行动空间的美化愿望，表现在他们的日常用品上，有时甚至体现在自己的身体上。

无论装饰的灵感是来自对大自然的模仿，还是抽象的几何图案的表达。平衡与对称一直是远古时期人类进行人体彩绘、纹身甚至疤痕装饰的原则。

人体的"美化"是人类最古老的习俗之一：要么装饰自己的身体，要么佩戴珠宝物件。无论如何，它的目的在于从审美角度，为人类的自然特征赋予更多的含义，以此来区分习俗、性别、民族、社会等级和年龄。在"美化"行为中，人体彩绘和纹身大多通过雕刻形成，它不仅阐述了人体永久和有机的装饰概念，同时也将装饰视为临时配件。无论如何，它们都成为人类出生、性别、信仰和社会关系的符号。

事实上，还存在其它以改变或损伤身体结构进行的"美化"行为，例如将装饰物嵌入耳垂或嘴唇，或者切割、修整牙齿。总而言之，这些人为改造的行为试图改变人类本身的自然外观。当然，除了美观因素以外，这些"美化"还具备一定的象征意义。例如所属的特定的社会等级或部落群体，或婚姻之前的青春时期，有时伤疤甚至传递成熟年龄的自豪感。

色彩象征主义在人体装饰中也具有极其重要的地位，丰富的色彩也赋予传统艺术以多样性。这些传统艺术的灵感往往来源于由神话装饰风格所激发的几何图案，红白色的之字形图案，具有消灾祛邪象征意义的图案，以及由折线、对角线、鱼刺符号形成的动物形象。

这些人体"美化"行为采用了不同的方法和技术，如今的创新技术使得美化过程并不存在具体装饰物，或使装饰物隐藏不被看见，并改变突兀的性别特征重塑审美原则，甚至希望借此永葆青春。

当装饰物为手工制品时，就装饰与材料和技术之间的关系而言，它具有丰富的可塑性和极强的叙述能力。这些都与产品的结构和功能紧密相关，又或直接叠加于其结构与功能之上。

相反，当装饰物为工业产品时，具有标准量产（铸造、压铸，旋转成型等）产品独有的特征。因此，无论产品表面是具有雕刻、凹槽等装饰图案，还是凹凸不平的三维结构，装饰都应提前成为产品基本形状的一部分，并与其美学特征保持一致。

无名 > 多少次我们驻足于商场橱窗前，静静欣赏着由螺丝、钉子、挂钩、铰链、钳子、胶水包装或电气开关等工具器材布置而成的"展览"？又或是那些剪刀、指甲刀、衣架、衣夹、发夹、订书机、刷子、梳子、剃刀、海绵、肥皂盒等等。事实上，无论是从基础技术的使用，还是遵循的机械物理原理，又或是人们的使用方式来看，这些物品都难以过时、难以被淘汰。

无名设计产品面向大众消费，不需要广告传播就已经深入我们日常生活，虽然无名但对我们影响深远。无论是习以为常的使用方式，还是一成不变的形式或包装，这些产品总是给我们一种熟悉感。即使我们试图增加它们的性能，但它们的外观也不会有太多的变化，我们只是会改进技术（增强耐受性、持久性、安全性），改变形式类型或是材料的选择。

产品流 > 相对于设计与实现建筑物的方式,构思与生产工业产品有着更为明显的自主权。工业产品大规模的生产,与其所处环境并没有特定的限制,因此适用于全球所有用户。相反,建筑物与城市元素结合,扎根于某处并阐释其当地特征。因此,产品与其背景,即产品与其构建的环境形成了一种值得深入探究的关系。我认为,正是这些产品流与空间形成的系统,给予我们一个了解当代社会的契机。

在设计建筑物或者规划部分城市的时候,我们需要满足特定的需求。事实上,这受到很多制约,比如可用的预算以及可能涉及的具体功能(学校、教堂、政府大楼、办公大楼、剧场、住宅,或特定的综合类型建筑)。

在城市地域规划方面也同样受到不同模式的制约。如住宅、服务体系和生产系统之间的关系模式,或城市基础设施之间的关系模式。总之,项目的预算和选址,涉及的标准与原型形成了设计师构思的基础,他们根据自身的技能提出了与项目相关的革新意见或创意想法。

相反,在工业产品设计过程中有着许多重要的因素。其中包括产品的功能特性,产品的生产流程,改善产品质量以达到增大市场份额目的和完善评估,甚至是产品自身所涉及的科学技术及其生命周期。然而,工业产品的生产逻辑是先提出产品概念,后形成用户需求、生产技术及其生产模式。因此,在什么都没有的情况下,工业产品的设计概念就应运而生了。

复杂的文化 > 随着发展中国家的民主逐渐形成,国内生产总值不断超越欧盟国家,设计领域正在经历从未有过的实验阶段。该现象在亚洲国家普遍存在(尤其是中国和印度),即使近年来拉美国家正在经历经济危机,但依旧不影响这一现象深入其中。从政治、经济、社会和文化的角度来探讨这个现象是很有趣的,但是我们决定把重点放在分析它是如何影响拉丁美洲的当代设计上。设计,即发明和制造产品的能力,换句话说,就是通过生产有形或无形的产品来满足人类过去和现在的所有需求。

拉丁美洲是一个广大的地区,鉴于自身在文化、传统、经济甚至使用语言(西班牙语和葡萄牙语)方面的不同,它试图建立南方共同市场。在如此辽阔且复杂的地区范围内分析设计表达,研究其中相同性与多样性的主题确实是十分有趣的。

在设计领域,即使对本土表达形式的兴趣正在逐渐增加,欧洲模式与北美模式的影响力也依旧是显而易见的。事实上,即使全球化趋势已无法避免,仍有越来越多的实验研究与生产活动,正试图恢复当地的传统文化与跨越整个大陆的复杂性文化之间的关系。

乌尔姆(Ulm)学院的教学方法影响了甚至依然影响着拉丁美洲各国学校的教育系统,比如1964年在里约热内卢创建的工业设计学院(ESDI)。这种影响为设计学院提供了教学规章与方式,但同时也创造了一个基于现代性伦理和风格的设计理念,这种理念适用于教育却并不真正符合当地的市场需求,也无法满足企业具体的发展模式。因此,说其阻碍了当地科学技术的发展也并不为过。

也许欧洲的设计传统,正是整个拉丁美洲设计领域的共同之处。由于当地文化受到全球化趋势的冲击,以及从八十年代开始并一直持续的拉美国家的变革,相同的设计基础文化开始逐渐"混合"。正因如此,这些国家地区的传统文化的全部潜力被激发出来,最终释放出了拉丁美洲大陆人民最真实最全面的独立性与创造力。

另一方面,该地区的多元文化和多民族特征促进了各国之间的相互影响。比如南方共同市场的建立,社会间的文化现象与表现形式的快速交流。这也就意味着所有的相同性与不同性正融入一个创新的混合形式之中。

无论如何,在新产品的设计与传统之间的关系中,功能转移这一有趣的现象是非常普遍的。也就是说,产品涉及的物质文化特征正从一种类型转向另一种类型。例如,办公用品的特征正转移到家居用品中;生活用品的特征正转移到市集物品或宗教用品中;日常用品的特征正转移到城市集体空间中。然而,这种方法倾向于重新诠释而不是重新生成产品。也就是说,并不是直接复制当地产品的形式、材料和传统工艺,而是对其传统历史进行重新解读。同时,转移这一概念也被充分利用在语言运用和形式研究方面:游行人群中的色彩,集市中的色彩,天然材料的色彩,传统服饰的色彩等等,都被转移到新的产品中;古老图腾中的几何图形、装饰纹样也都成为新的设计方案的参考准则。

在拉丁美洲国家,至少在其中GDP增长较快的国家当中,设计方法的另一个重要特征则是自我生产模式(self-production)。这是非常有趣的,因为它由一个相同的因素保留衍生出了多样性特征。所谓多样性特征,主要体现在产品本身,与其材料、地域历史、文化和经济、社会和人类学背景息息相关。所谓相同因素指的就是自我生产模式,一个根植于但同时又超越传统工艺的系统,它并不涉及历史发展中的先进的工业生产模式,而是以市场的大量生产需求为基础并发挥自身最大的可能性。事实上,设计师使用现有材料,借助技术和工具来生产自己设计的作品,因此,设计师就此成为经营者。但是,他所面对的市场是由数量所限制的,即便如此,这种模式依旧允许设计师提出新的设计方案,生产新的产品,形成一种不断激发创意的良性循环。这种模式在生产技术与产品类型方面,不同于传统的生产模式;在生产阶段与管理执行方面,也不同于工业生产。

自我生产模式开始逐渐关注可再生材料,这是一个普遍趋势。坎帕纳兄弟(Fratelli Campana)无疑是该领域研究的先驱者;同样的,一些像Notechdesign的年轻设计团体甚至以此为主题确立设计宣言。这是一个有趣的研究,在设计师方面,它恢复了手工技能及创造力的维度。在产品方面,除了固有的环保象征意义以外,它还赋予其艺术特性及刺激感官的能力。

显然,自我生产模式是带有争议的。许多人认为它阻碍了国家工业体系的发展,削弱了与先进工业国家的创新竞争力。

然而毋庸置疑的是,无论是在产品设计、时尚设计、视觉传达设计,还是建筑设计与室内设计方面,通过设计在拉丁美洲大陆进行的实验研究是充满原创和创新精神的。

§
[3]

摘 录

引 言

这些年来,我倡导发起了一些出版工作,如城市规划、设计与建筑技术系的刊物,又或者是我领导院系的教师讲座的文集。

辩论最为激烈的教授团队参与了这项工作,他们的一些文章阐述了与建筑和设计相关的主题。当我再次阅读了他们的大量文章时,我摘选了部分内容,用对照法进行叙述,以摘录的形式来对本书进行完善和收尾。为了丰富本书内容,我按照以往的经验选择文章选段来探讨共同感兴趣的主题来支持阐述"罗马大学的设计"(design alla Sapienza)。事实上,这些主题非常难得地将众多思想汇聚一起,以此来加深我们作为与设计文化相关的教师和研究人员的思考。随着时间的推移,它们还进一步加强了相互之间的影响。这部分所引用的文献围绕特定的主题进行汇总,而这些主题则成为阅读完整内容的关键词汇。引用的参考文献分别为来源于:文森佐·克里斯达洛(Vincenzo Cristallo),范德利克·达·伐柯(Federica Dal Falco),洛雷达娜·蒂鲁克(Loredana Di Lucchio),罗勒佐·尹百斯(Lorenzo Imbesi),萨布里娜·鲁奇百洛(Sabrina Lucibello), 拉依孟达·利奇尼(Raimonda Riccini)。

设计命名

诠释 > 谈到生产范畴,我们会想到产品设计、时尚设计、食品设计、交通工具设计、照明设计以及建筑设计、城市规划设计。谈到生产阶段中的具体活动,我们会想到交互设计、传达设计、生态设计、零售设计以及品牌设计。后来,我们每次谈到设计,即使是相差甚远的领域,也会出现某种设计形式,比如音响设计、花艺设计、婚礼设计等。在这样一个复杂多变的环境中,设计到底是什么?它的作用又是什么呢?

每当我们想对设计进行思考时,总是不可避免地会遇到两大问题。第一,这不是新的问题,是要对设计实践进行最恰当的定义;第二,这是较近期的问题(更加基础),是要了解设计与生产、文化及社会体系之间的关系。两个问题中的第一个,我们可以总结为:"设计可以做什么?"这是设计学科诞生之后就一直存在的问题。只要我们想一想工艺美术运动开启的对立状态;或是德意志制造联盟的推广普及;又或是包豪斯成立时的宣言;直到二十世纪下半叶引发的辩论,试图通过设计产品来描绘设计的社会职责,以达到社会的认同。我们只是不曾知晓,这场辩论发生在意大利,那时还衍生出许多有价值的观点。

吉洛·多福斯(Gillo Dorfles)提出了设计的第一个定义(正好在1958年),他写到工业设计是"通过工业化方法和系统对产品进行大规模生产的特定类别设计,技术方面一开始就与美学元素结合"。但是吉洛·多福斯(Gillo Dorfles)所描述的这种"工业化"生产方式完全针对产品外观,并没有对工业设计的其它方面进行阐述。另一个由恩佐·弗拉戴利(Enzo Frateili)提出的定义对此进行了完善,"不同的学科交汇,使得产品具有生产—分配—消费的完整周期"。

因此,根据恩佐·弗拉戴利(Enzo Frateili)的说法,作为设计活动的成果必须具备两个特征:由于设计的工业和工程性,所以产品应当具备量产特性;由于产品自身的创造性和涉及的社会文化,所以应当具备美学属性。其中最合适的概念还是托马斯·马尔多纳(Tomàs Maldonado)几年之前提出的,他认为工业设计(再次,不是设计)是"协调,整合和阐明所有这些因素以某种方式参与构成产品形态的过程。更确切地说,是将产品的使用、实现与个人或社会消费(功能、象征或文化因素),与产品的生产(技术/经济,技术/构成,技术/系统,技术/生产,技术/分配因素)相结合"。设计的本质是操作从外部(社会与企业)接收的信息;重新组合融入新的形式;在近乎连续的周期中再次将它们转移到外部(如果我们不局限于考虑单个设计者的行为)。那么显而易见的是,新信息技术在解构获取和传播知识的相应过程中,改变了社会—设计—企业—社会的循环,使其大体上呈现扁平化。如今在信息技术领域,尤其是自开放源代码(open-source)的出现,设计在许多其它文化和生产环境中被阐述与提及。

人工造物 > 在《人工造物》中,埃齐奥·曼齐尼(Ezio Manzini)写道:"产品可以分为两大种类:延伸性产品,即从生物学角度增大人类可能性的工具;符号产品,即具备的模糊的含义成为自身构成的一部分的产品。如今,这两种说法已经无法满足。新的产品家族不断出现,它们执行复杂的功能,比如对大脑和感官活动进行多重刺激,这也许就是第三类产品。上世纪九十年代开始对教父产品、设计明星和品牌认同文化展现诸多崇拜。"二十一世纪，随着现有产品种类呈指数级增长，这些产品种类之间的差异似乎越来越小。新类型的增长，并不代表着它们与之前的类型发生了实质性的变化，而只是自然而然地习惯性地去改变。这种现象在统一功能和语言的过程中，涉及的差异微不足道，实际上并没有带来任何有价值的创新。持续变化的产品不断地影响着我们的生活，使得我们所处的环境越发具有流动性。然而，即使这些产品看似新颖，但整体环境并没有发生可以称得上进化的改变。

正是在那个年代，近代最有影响力的认知科学家之一唐纳德·诺曼（Donald Norman）在反思设计的角色时，在他的《日常事物中的设计》（The Design of Everyday Things）中指出，设计只不过是在关注外观形式，有时过多的处理反而让产品变得无用。

即便如此，当设计在这种情况下不断发展时，在其它知识领域（制造领域）中，我们意识到社会正在形成一种新的变化。社会文化系统不再由物质财富来决定，而是由无形财富，即技能，尤其是个人、团体和整个社会的智慧来决定。

"知识经济"时代所提出的理论，投资的新资本是知识资本和声誉资本，即知识本身从"给定因素"转变为"构建要素"。以产品、服务和媒体的形式，知识正从生产地点转移到消费地点，并且能够不断地产生并再生新的知识，逐渐成为行业内的决定因素。

全新的场景就此诞生，其中新的组织模式（企业、机构或是社会团体）不再是单一的、同质的、可复制的。相反，它们是能够对各种刺激（社会，文化和政治）作出快速反应的可变模式，因此具有高度灵活性和适应性。因此，显而易见的是，这些不断变化的组织模式，要求其中的参与者也必须具有同样的能力。

物品传记 > 无论很快消失还是长久存活，每一件产品都拥有自己的生命周期，我们称之为"产品的传记"。

在考古学领域，从哲学的角度来看，这种自然界生命周期的说法时常被运用在物品循环中。根据这个观点，我们可以沿着两个极端进行物品路径的研究：从最初始的吸引到衰败时的抗拒。通常，一旦快速消费使得物品异化，它们就会消失，或逐渐变得无用、陈旧与怪异。之后偶然间被发现得以重生，人们就开始好奇它们的历史。大量的物品被当作遗产传递，而另一些破损的物品则被扔进垃圾场，留在地下室、杂货店或古董店。并不是所有物品被处理的原因都能被解释，不过通常是因为失去了使用功能。"那些幸存且一直被使用的物品往往是经历了一个或多个生命轮回。"总而言之，物品随着环境的变化而改变，如果它们能够幸存下来，那么它们就会向我们传递信号，向我们诉说它们过去的多重生命轮回。这些物品的经历证明了随着时间的推移，再次的使用功能总是以原始结构为基础的：通过标记它们的特征能够定义一个或多个时期。通常，大多人更加关注那些特别重要的物品，他们极少注意到那些普通无名的家居用品，比如熨斗、扫帚、剪刀、肥皂等等。最近，弗朗西斯·里戈蒂（Francesca Rigotti）对它们进行了研究，将它们作为真正的主角放置在历史的长河之中，进一步赋予重复的日常行为以更高的价值，因此将这些小物件从使用功能中释放出来，让它们开始讲述自己的故事。比如，用来阅读的报纸，此刻却用来点燃炉火。事实上，这些隐藏其中的所有含义要比简单的用途复杂得多。自古以来，人们普遍认为，物品是象征符号的载体，这些象征意义往往来源于物品本身而非其技术因素。从这个意义上说，这些物体被认为是"活跃的存在"，因为它们由难以量化的复杂关系组成，并逐渐成为"人类宇宙的一部分"。根据这个概念，设计应该被看作是一种随着时间的推移而进行的活动，不仅仅是在为平淡的实践服务，而且是为了传达仪式和情感因素。随着远程信息技术的革新和新的设计实践的兴起，这种动态通过新方法的实施变得更加突出。如今，交叉学科似乎已经成为实现创新的唯一途径，它定义新兴生活模式和与之对应的风格，通常也融入类似的混合技术，试图将现实生活与虚拟世界相结合。但是，也存在许多遵循线性进化永不衰败的物品，例如雨伞总是被人们所需要。这些物品的价值并不在于其外观装饰性，也不在于涉及的机械智慧，而在于其形式与功能之间的紧密且完美的结合。物品被遗忘又或进化，它们的存在间断性地将信息传递给我们，它们用一个时代的思想、品味和风格来记录个人的记忆，甚至记录了一段文化和历史，一段神话和一个事件。

创造力 —— 高频词汇 > 在颠覆性的创新时代，打破常规思维是必不可少的。即使人文文化（人文精神）在大学教育中逐渐落寞，但它也是科学家必须具备的。恩波利（Da Empoli）认为，在新的人文主义中，能够让大脑中艺术和技术两个半球同时运作的人，也就是设计师。但这不过是陈词滥调，也有些过于关注传奇且短暂的创造力。

乔·彭蒂（Gio Ponti）在 1957 年做了这样的表述："原来的观点不算；实际上，原来的想法并不存在；想法被接收并重新表达；想法是我们思维世界的一部分，无论是从过去到现在还是未来；人们总说'我有一个想法'，而不是我'创造'一个想法：从词源上来说，发明意味着发现，而不是创造。"在我的这场讲座中，"创造力"这个词必不可少。它就像是一个朦胧的精灵，破坏了从事设计的人的梦想和确定性。它好像是一个无法被看见的灵魂，你永远不会知道它的真实身份。它是如此不确切，不禁让人感到折磨。我们对它过于肤浅的了解，有时甚至让人感到受伤。媒体寻求简单明了的叙述方式来阐述设计，从事设计的人员总是颂扬浪漫主义文化，他们一直将"设计"与"创造力"捆绑在一起，给人一种强烈的不适感。如果可以的话，我希望暂停一下：五年之内禁止使用它，看看这会带来什么样的影响。事实上，它现在已经开始被其它词语取代，尤其是在通信领域中，被更加迷人的"智能"所取代。尽管如此，"创造力"带来的财富依然存在于各种文学作品中，与它相关的"神话"主宰了许多人的想象。在我等着"创造力"被暂停的期间，也就是几个月前，我在罗马地铁的地下通道中看到了一个意想不到的方案，我想这可以成为让师生们进行一场辩论的契机。这是一则跟我们拥有相同专业并有所竞争的学校的广告，它占据了 120×180 厘米大小的广告位，上面赫然写着："如果没有设计，创造力什么都不是！"但是这些名言，即使像奥斯卡·王尔德（Oscar Wilde）的言论那样尖锐和有文化，也终究不会长久。想要在教育领域内（我们的职责）改变这一想法，需要很多的努力。这样一来，"创造力"一词"因为被过多使用，逐渐进入了一种空洞的表达形式。事实上，它的概念并没有什么实质内容，是完全通用的，因而才得以广泛传播。"

因此，我们必须建立一道屏障，来消除"创造力"是设计方案来源这一蛊惑。抵制设计师的特定形象，设计师并不是拥有天才般的才能，灵感一闪就能得到解决方案。同时，要反对教学方面存在于概念上的偏差，比如"服务"、"战略"，它们并不是疲惫不堪的设计师用来疗伤的咒语。事实上，每个人都知道"创造力"是什么，但是如果不翻阅词典（词典解释：创造性能力，产生新的想法）的话，没有人知道它的精确含义。

著名的出谜题者斯坦法诺·巴尔泰佐吉（Stefano Bartezzaghi）认同这一观点，他认为这样的高频词模糊且不确切。在某些情况下，在技术方面词语本身就是一个矛盾体，它存在的意义就是迷惑那些从事设计的人。那么，我们不能接受将创造力以研究的形式，应用于设计教学领域。也就是说，作为设计学院中的一员，我们应当不惧艰险，向学生们解释什么是创造力，什么不是创造力，应当如何分辨、如何规避它。在教学过程中我们应该采用一种清晰而不是隐晦难懂的形式，否则则与我们的目的背道而驰。如果我们真的要踏上这条艰险的教学之路，那么有本书不得不读：《科学与方法》，它是由亨利·庞加莱（Henri Poincaré）（法国数学家、物理学家、天文学家、哲学家和科学家）在1908年撰写完成的。他对"创造力"的观点简单但却值得被采纳。他认为创造力是"将现有元素组合生成新的对人类有用的物质的能力"，并强调决定新物质实用性的直观检验是它的美。如果想要评价这种美，需要具备一定的能力和情感感知力：它是与优雅、和谐、经济相关的符号，是对最终目标的一种回应。亨利·庞加莱（Henri Poincaré）提出的定义是丰富、清晰、具体且普遍存在的："它适用于科学、艺术、技术领域。"他的思考也直接与教育相关：创造力建立在打破规则的前提下，因为首先需要掌握知识，没有对应技能就无法继续发展。因此，富有创造性特征的人应当认识到掌握技术的重要性，拥有好奇心，对秩序和成功（不是经济方面）有所需求，人格独立，永不满足且具备批判精神。

设计的伦理 > 作为设计师、设计哲学家、文化人类学家和教师的维克多·帕帕奈克（Victor Papanek）严厉地阐述道："在所有的行业中，最具破坏性的是工业设计。也许没有比它更虚假的专业了。"他用精确而戏剧化的语言攻击了工业设计师（他也把他们定义为危险的种族），并预言工业生产的世界应当更具社会性和环保性，因此设计伦理也必不可少。这也就意味着设计应当担负社会责任，设计师必须了解每个设计思想和行为对人类及其所处的社会产生的后果。

维克多·帕帕奈克（Victor Papanek）确信，在大规模生产的黄金时代，所有事物都需要进行规划，"设计已经成为人类塑造所处环境当中最强大的工具。这也意味着设计师需要具备高度的社会与道德责任。

摘录 _ 设计命名

因此，设计必须成为能满足人类真实需求的新型的、高度创造性的跨学科工具。设计也必须更加注重研究，不能让设计不好的产品和建筑污染地球。"将工业设计视为一种能够具体满足人类需求的工具，这是一个明显矛盾的说辞。因为，维克多·帕帕奈克（Victor Papanek）的意识基础是批判、限制、界定工业设计师的角色与工作，但最终却将其释放出来。从最初的观点出发，不考虑所需工具与内容，设计是人类每一项活动的基础，人类借助这些活动有意识地改变着周边环境并赋予其重要的秩序。维克多·帕帕奈克（Victor Papanek）认为，之所以使用"重要的秩序"这一术语，是因为含义过于丰富的表达，如美丽、丑陋、优雅、恶心、现实、黑暗、抽象与可爱，并不能阐述真正的含义。总而言之，维克多·帕帕奈克（Victor Papanek）所有对于伦理的阐述都处于"设计意义"的层面：设计是道德的，因为它是一个帮助世界的工具，因此我们使用它，也让别人使用它，并最终在学校教授它。

意大利设计大师之一恩佐·马里（Enzo Mari）为了避免误解，使用"伦理"一词本身来阐述伦理。但是他首先说："乌托邦是不存在的幸福之岛，它是不可能存在的。我是法国革命之子，我们都是天主教徒或犹太人、马克思主义者或资本家，'平等'（egalite）是我的信仰，但我并不认为这可以实现。即便如此，我依旧愿意相信，我们可以通过设计实现平等。""乌托邦"能和"伦理"相提并论吗？恩佐·马里（Enzo Mari）就这么做了，还是以一种让人信服的方式。他说道，"设计伦理就是设计目标，就好像是希波克拉底誓言。我的意思是所有的设计师应当记住每个医生毕业那天的誓言，我们可以将其简化。即首要的任务是拯救病人，但也不要忘记把乌托邦（一种新的可能的模式）作为伦理的扶手来引入伦理改革。"这种说法是不是让人摸不着头脑？这虽然看起来更像是贾尼·罗大里（Gianni Rodari）的寓言神话，而不是恩佐·马里（Enzo Mari）的说话风格。但是这就是他对自己想法做出的解释，也是对这个主题的直接阐述："当人们问我，在我认识的人当中谁是最棒的设计师？我总是这样回答：一位在树林里种植栗子的老人家。靠水果充饥，用木头取暖或制作凳子，夏天在树荫下乘凉，我们都知道这些明显不足以维持生活。他种这些不是为了自己，而是他的子孙后代。"为了更好地平衡这些想法，我向阿切勒·卡斯蒂廖尼（Achille Castiglioni）寻求帮助，透过他的作品我们感受到了其中的设计文明，从文明的本性中我们也可以看到伦理，事实上，它与阿切勒·卡斯蒂廖尼（Achille Castiglioni）的思想紧密相连。设计伦理可以被视作设计好坏的标准，也是选择事物的理由。在赛尔吉·波拉诺（Sergio Polano）的记忆中，阿切勒·卡斯蒂廖尼（Achille Castiglioni）"似乎充分地认识到产品的流动是永不静止的，正如乔治·库布勒（George Kubler）在他的《时间的形状》中提到的一样：今天存在的事物都是对过去的临摹或演变，事实上它们从来没有真正改变直至回归初始。"不约而同的是，布鲁诺·穆纳里（Bruno Murari）也认为，当有人说"这件事我也会做"的时候，是指这种方式显而易见并不会给简单的事物带来价值，他们最多可以说他们可以重新再做，要不他们早就做了。正如历史学家无法清楚地区分不同性质的事件，如果不考虑"复制品"的传统，即使是再谨慎的设计师也无法设计出下一代"复制品"。这种说法，更加凸显了阿切勒·卡斯蒂廖尼（Achille Castiglioni）对日常工具的喜爱之情，这些物件沉淀了人类的智慧，蕴含了恰当理性的制作逻辑。阿切勒·卡斯蒂廖尼（Achille Castiglioni）在其作品中展示的伦理价值清晰明了，他并不是在寻找一种无名的社会主义情感，而是试图将主观设计与客观生产相结合。

对于摆脱教条约束的理性主义者阿切勒·卡斯蒂廖尼（Achille Castiglioni）来说，有必要对产品内部的特性进行研究以便持续探究其真正的含义。也正如赛尔吉·波拉诺（Sergio Polano）所言，即使不断地追求"减法甚于加法"的方式，我们同样可以看出产品的质量依旧具有一定的价值与意义。

现代与后现代设计危机 > 设计一词的主旨标志了产品设计的整体理论。如今，设计覆盖了全球范围内有形或无形的产品，但是当设计被频频提起时，它的含义却并不那么明确。这些年来，对设计意义的过多控制使得"设计"成为一个品牌标识，人们再无法了解它的理由与起源。我认为，我们对这个学科产生了致命的误解，也就是说，这个词汇成为了理解当代设计中现代理论的障碍。这个词汇意义的恶化，在我看来是由一个坚实且重要的原因引起的，即设计的后现代主义倾向。这不是天生的，而是随着时间的推移环境的改变产生的。工业设计自诞生以来，其担负的历史任务的限制始终都与现代主义原则紧密相关。随着设计之后的定义在表达上更加自由与包容，它成为后现代主义的代名词。因为当代设计的基础思想与前提文化已经发生了改变，所以它在使用产品领域已经无法再产生新的变化。工业设计被设计取代，成为任何根据典型的后现代主义模式从事设计活动的集合，它包含了所有的可能性。设计作为后现代主义概念，改变其形式与观点，并运用其煽动性手法及伦理与援救的立场，于是商品学与形式主义应运而生。对大多数人而言，利用可见及可复制的手法，设计让混合概念更容易被理解。如今，每天都有新的现代产品诞生，使得"混合"能够超越所有后工业的预测。事实上，工业与非工业产品，是设计领域长久以来一直存在的矛盾，意识文化与工业实践的支持者始终各执一词。然而只有设计逐渐替代工业设计，才能对其进行真正的区分。然而，在后现代主义的使命中，设计已经成为意识文化与工业实践共存的重要因素，同时也在其形式、模式和新型文化方面为众多实验研究提供空间。

后现代主义是含糊暧昧的，归根结底是因为它内部保留着想要推翻它的现代主义。然而后现代主义的衰落则源自于一种简单的艺术流派，它不断发展演变直至具备政治及社会意义，但是它们从来不是真正的对世界的预先构想。它的意识形态就此瓦解，因此每个对后现代主义的阐述都同样可行。这个愿景虽然提出了一个宽容多元化的民主社会模式，但是终究超越了客观性的想象，而且对事实进行诠释的重要性高于事实本身。

真实的世界由最初童话阶段进入现实阶段，但现实世界中的民众主义媒体能够使人相信任何事物。然而颠覆后现代主义最极端的理论就是民众主义，也就是说，事实与其诠释之间的位置正发生着改变。

在设计领域，我们可以将这种混乱视为一件神奇的工具，它见证了今天的一切都是设计。在某些情况下这也许是真的，但无论如何，并不是说所有设计的事物都是设计的。后现代主义和与其对抗的民众主义，以及由事实与其诠释产生的混乱的想法，就像是这些年来不断被提出的设计与建筑之间的矛盾。

设计文化中的材料 > 许多历史性和批判性的著作已经将注意力集中在意大利设计文化的具体特征上，记录了它如何成功地超越了传统的技术文化从而走上了一条创新之路，并在语言、形式、功能、生产质量与材料创新之间取得平衡。

意大利设计的这一特点不断地改良更新，在某些特殊时期表现得尤为明显。

上世纪五十年代，意大利在设计初期便形成了自身的独特性，即将实验研究与材料相结合。随着在工业领域莫普纶等规聚丙烯纤维（Moplen）的广泛应用，卡特尔（Kartell）公司就此诞生。借助缤纷的色彩、轻盈的材质和民主的风格，卡特尔（Kartell）迅速为日常生活带来一场变革。

随之而来的七十年代，安德列亚·布兰茨（Andrea Branzi）提出的"原始设计"将注意力转向材料的柔软性和感官刺激方面，将设计能力扩展到产品表面的再造上。正如我们所理解的，"材料设计"就此诞生。即便是在以 Montefibre 公司为首的商业世界，也将工业战略转向材料，潜心研究材料的色彩、装饰（特别是织物）、表面、光线及声音。

八十年代后期设计迎来了创新技术的飞速发展，走进了一切皆有可能的世界。在这里，具有非凡性能的材料突破了前所未有的极限，成为技术含量与象征意义的代名词。与此同时，其它领域的材料也纷纷进入设计行业。

这样的情况持续至九十年代，在此期间谈及材料就意味着需要具备对复杂事物的认知与管理能力，因为材料不再是设计师或技术人员可以单独进行处理的简单事物。九十年代也对未来充满了预测，认为未来是完全小型化和非物质化的世界，值得庆幸的是这些预言未能实现。直至今日，设计方法和策略依旧处于不断的改变当中。设计师试图发挥材料的所有潜力，使一切变得可行，同时也试图重新定义安东尼奥·佩特里略（Antonio Petrillo）所说的"新的人为世界的秩序"。

这样一来，新型技术不再仅限于延伸并改变材料的外观属性并赋予其个性化特征，而是作为改变感官特征的工具，从而影响我们的感官反应。正如纳米技术在宏观、微观甚至纳米层面进行操作，使得产品在表面上极其简单，而实际内部愈发复杂。

这是一场真正的设计革命，我们称之为"超设计"。在操作材料的同时对形式和功能进行研究，为超级材料开辟了无限的空间，让它们更加轻盈，更富感官性，性能更强，更加环保，更加神奇以及也许更加"鲜活"。材料的创新可以以两种方式进行：如果说一方面是超级大国提供的超级材料，使用技术科学促进其发展；那么另一方面是我们所说利用"创造性"方法，即使缺乏原材料和资金的投入，也能够在产品和材料层面实现创新。意大利设计尤其如此，在资金和原材料匮乏的情况下，仍然创造性地使用了这些材料，并通过新产品宣扬了意大利的创新设计。

所以，创造性思维可以想象那些理论存在但实际不存在的事物，同时促进艺术与技术领域的研究试验。

产品、材料和使用方式的改变，引发了大大小小的革新，赋予了生活千变万化的可能性。借助纳米技术，人们可以在分子、原子层面对材料及其构成进行修改，也正因如此，可以模仿大自然将特定属性植入材料内部。比如莲花效应，即莲花出淤泥而不染的自我清洁能力；又或者壁虎特征，即依靠爪子下方特殊的结构形态停留于各类墙体表面的附着力。

举一些有趣的例子。受莲花效应的启发，杰尼亚集团 (Ermenegildo Zegna) 研发生产了一款名为 traveller 的防污面料，它能够在不改变织物性能及触感的情况下进行自我清洁。另外，利用纳米技术改变自然材料生物 LED 的遗传物质，赋予它们丰富的色彩，也许在未来还可以借助这一技术将树木变成路灯。使用纳米色彩可以在不需要合成颜料的情况下进行上色，同样地，可以通过模仿蝴蝶或其它昆虫的翅膀结构来制造色彩。

众所周知，蝴蝶的色彩得益于翅膀中的微型结构。光线照射在翅膀表面，通过折射与反射原理产生了缤纷的色彩。对自然光子结构的模拟开辟了一个有趣的场景，同时构建完善了能够防止伪造纸币、信用卡和其它文件的工业系统。在仿生学领域的宝贵财富中，我们不仅可以模拟和谐优美的自然形态，还可以效仿数不尽的有机结构，甚至是自然有效的循环系统。依据生态循环系统的原理，将废料转化为营养物质，新的可持续发展方案由此诞生。正是由于众多设计师的推广，可持续发展概念得到了更好的普及。其中，2008 年，设计师马蒂尼·雷阿奴 (Mathieu Lehanneur)、莉薇·科恩 (Revital Cohen) 与苏珊娜·索尔斯 (Susana Soares) 共同参与了由宝拉·安东尼利 (Paola Antonelli) 策划的位于纽约现代艺术博物馆 (MOMA) 的《设计与弹性思维》(Design and elastic mind) 展。在展览中，他们共同呈现的 SymbioticA 作品结合了不同的生物，通过生物之间的交互活动产生了感性的色彩。同期展出的还有设计师杰特·凡·阿巴马 (Jelte van Abbema) 名为《共生》(Symbiosis) 的作品，他把字模做成了细菌和细胞的培养皿并将大肠杆菌置于广告牌中，随着时间的改变、微生物的死亡与繁衍，整幅作品也会随之改变形态。

还有许多有趣的试验，设计师试图尝试着去实现类似于细胞结构的人工材料。比如木醋杆菌 (xylinum)，与生产它的细菌同名。用葡萄糖喂养木醋杆菌，它便能产生百分之百可降解的细胞纤维结构，与此同时还可以通过改变细菌的遗传物质改变材料的性质。

技术创新以一种反向模式带我们回归自然，事实上，我们可以从多个方面进行模仿。我们不但可以临摹大自然的形式（复杂的生产流程与信息化网格控制使得形式更加有机），还可以效仿大自然的行为（更加自然活泼并且能够自给自足），甚至于我们可以学习大自然永无止境的多样性。

§
[3]
p.107

设计中的标准

称之为标准 > 安德列亚·布兰茨（Andrea Branzi）是设计发展史的倡导者之一，他用一个极其形象的说辞对其进行了描述：从史前时期到当代社会，在转化过程中存在单一的产品链，它们环环相扣，在人类历史及传统文化的养料中不断前行（也许是不停地舞蹈）。大量共存的产品类型老旧更替，形式推陈出新，这些持续不断的变化促使着我们的想象不断地发展。试想一下，如果我们采用人类进化史来学习物品的生命，那么我们可以将"灭绝的"、"存活的"、"进化发展的"产品进行分类展示，从最原始的一端直至完全进化的另一端。

在工业产品的生产模式中，它们被称为标准。所谓标准，是指在生产过程之前已经完成了的，而且在之后的过程中不能再对其进行操作与改变的原型，这一原型使得所有产品保持一致。这个经典的定义对产品进行了明确的划分，比如手工制品与工业产品，以及产品是否借助机器进行生产等。同时它也包含了混合产品的概念，即使用混合技术生产的产品，比如工业化生产的木制家具最终以手工形式完善其外形。但如今我们重新对产品进行分类时，就不得不考虑一个基本主题：设计及其历史的关系。雷纳托·德·夫斯克（Renato De Fusco）最近提出了产品的分类体系，该体系基于工业设计的自主起源，并非简单地延伸过往，而是一个与古老谱系相关的复杂现象。随着时间的推移，虽然产品的形式发生变化但无论是在手工还是工业层面，其本质却保持不变。

从工业设计的普遍现象（从设计到消费）看来，即使它形成了统一的概念，却依然被划分至不同的领域，涉及更多更广的商品种类与材料。

产品特征 > 建筑史学家彼得·卡尔（Peter Carl）说，自启蒙运动之后，也许我们真的处在类型学理论的第四个阶段。虽然彼得·卡尔（Peter Carl）说的是建筑领域，但对于设计学来说，这种说法也并不是不合适。哪怕只是大概的描述，我依旧试图回想"标准"这一理论的历史发展及其在产品设计中的作用。面对产品物质世界的明显混乱，技术存在对思考、设计、实现乃至产品（新或旧）的使用方式进行了彻底的改变。我们不得不承认，在二十世纪推翻的众多理念中，"标准"的概念也正逐步被抛弃。建立现代物质世界的产品从没有像今天一样如此颠覆。我们也从没有注意过，电子及集合技术、智能材料以及设计软件（其中也包括开放源代码）的广泛使用，使得形式不再是产品世界的中心，对形式及功能的探讨辩论也随之消失。

我们不禁开始思考，在设计当中"标准"一词到底有什么意义，如此一来我们不得不考虑它的两面性。一方面，标准是现代（标准化）产品设计与制造的基本典范；另一方面，标准是在社会场景（类型学）中识别产品的基本要素。因此，标准可以是理论的也可以是实用的，可以是发展的也可以是生产的，可以是与产品有关的也可以是与设计相关的。只要想到这里，就足以理解它对设计文化的重要性。尽管类型学并没有在设计历史及其理论中起到主导作用，但至少在我看来，它依旧与设计、生产和消费活动紧密相联。然而在建筑领域，即使历史与理论研究不断产生变化，标准及其作用始终处于重要位置。虽然设计文化（尤其是意大利设计文化）继承了建筑学的这一观点，但是却没能深入发展。事实上，标准、类型学、标准化这些术语，作为工业化成果和信息工具，已经迅速地融入现代主义概念之中。我们甚至常常忘记，产品类型在物质文化与使用功能（现实与虚拟）中的主要作用。毋庸置疑的是，类型学与标准化模式是物质世界生产与再造的主要工具，同时它也赋予人类处于其中的重要意义。回顾过往，以人种志学者的视角，我们可以将物质文化中的人工造物视为（时而快，时而慢的）发展过程中的变量与不变。举例说明，如碗、锤子、刀、凳子、篮子和花瓶之类的原型物品一定程度地代表了类型学的原始目录，随着不同时期不同地域的社会发展带来的多元文化，目录中的产品类型也不断丰富起来。但是更为复杂的是，在现代主义初期伴随着 homo faber 的原始操作，工业革命诞生。随之而来的是大量的技术产品和新生产品，它们进一步地成为了现代物质文化的主力军。

标准与现代运动 > 在早期现代主义的战略之中，标准化是大型工业化生产必须达到的要求，而现代运动的领导者则充分认识到这一点。古斯塔夫·普拉茨（Gustav Platz）就是其中之一，1927年，他在《新时代的建筑》中综合了二十世纪前三十年间的设计思维及相关生产，他表达的观点也得到了大多数人的认同。古斯塔夫·普拉茨（Gustav Platz）说道："大型化生产必须遵循标准与规范。无论是织布还是家用工具，无论是简单的物件还是复杂的建筑元件，所有的产品都需要借助机器进行生产。标准的定义不仅是机械化生产流程的成果，也来源于典型需求的不断重复，从而构成了经济正确的政治社会对标准的需求。1955年，瓦尔特·格罗皮乌斯（Walter Gropius）也在其文章强调应当将发展标准视为重要的主题之一。他说，"创建日常用品的标准是社会需求之一。标准化产品并不是我们这个时代的发明，我们只是改变了具体的生产模式。即使在今天，标准依旧代表了最高水平的文明，对完美标准的追求是将人类及发生在人类身上的事物进行分离。"在资本主义初级阶段，标准化应用于规范产品及其生产秩序，同时也一定程度地限制了产品类型的快速增长。如今市场的生产模式已经不像福特时期一般严厉，尤其是随着技术的飞速发展（机器人化），组织逻辑（及时化生产）发生了改变，市场需求也就是消费者得到了越来越多的关注。不需要回顾从丰田主义以来的所有改革，我们就可以肯定的说，我们正处于"制造转型期"，这完全得益于广泛普及的技术和日益突出的消费者、设计师以及生产者形象。

[3]
p.109

新的范式

开放源代码时代的设计 > 随着开放源代码（open-source）及自由软件（free-software）的普及，模式范式之间的界限越来越模糊。如今，单一的专业性让位于全方位的跨学科领域，设计师不再是独立而高傲的艺术家，设计过程也进入创意阶段，进而建立了一个开放互助的模式。面对这些挑战，设计及标准化生产帮助我们找到解决方案。在此，方法的正确与否并不重要，而是应当共同讨论这些设计现象及其问题、方案和涉及的背景环境。提到设计，我们就会联想到流动性、连接和转变。设计从垂直化向扁平化发展，从层层下放转向层层上传，从专利保护到开放源代码分享，从独立特征到特点遍布。同时，设计也带来了许多新的意义：为理性世界带来的感性价值；暧昧的价值与丰富的感官体验；现实与虚拟的模糊界限；超理性状态的回归，即关注产品与其主体之间的关系。当我们谈到关系性产品，正如唐纳德·温尼克特（Donald Winnicott）提出的过渡关系理论，应强调自体形成时环境的重要性。因此，不能简单地设计单独的产品，也应当考虑产品与其所处环境、与用户之间的关系，其服务体系中的多点触碰以及转变过程，等等。设计蔓延至所谓的间隙空间，为新的范式、新的知识领域以及新的思考方式打开了一扇大门。1958 年，为了研究工业基本操作的原型形式，恩佐·马里（Enzo Mari）为 Danese 设计了限量版的"putrella"餐桌托盘。他依托现成品艺术（Ready Made）背景，通过分解并重新制定功能，将一段工字钢的两端弯起形成托盘造型。这一产品具有建筑工地特有的半加工模式，对于资产阶级家庭而言，接纳它无疑是一个很大的挑战。它同时又以其独特的两端弯曲的形式调侃了工字钢的功能与耐受力。相对于罗兰·巴特（Roland Barthes）对语言的分析，以及同时代的设计师来说，恩佐·马里（Enzo Mari）借助形式和标准，向功能规则与象征意义发起了挑战，像黑客一样汇集所有进行标准生产的技术因素，将其重新操作形成一种新的含义，以灵活的方式再次宣扬了生产的标准化。

多年之后历史再现，它成为一个开放且永无止境的过程。2014 年，罗内恩·卡杜辛（Ronen Kadushin）以开放源代码（open-source）的形式对"putrella"进行再设计，发布了激光加工"putrella"的电子文件，使得用户可以直接生产组装托盘。他在预先开孔的金属板上切割出工字钢外形，然后进行拼接组装，此时的产品以二维模式呈现，不过也失去了最初产品具备的建筑材料的坚固特性。另外，吉尔贝·西蒙东（Gilbert Simondon）也认为，技术在平衡原型形式与物质、社会尺度之间的关系中具有举足轻重的作用，同时，技术也有助于新类型的增加，为物质世界带来了巨大的财富。

桌面制造设计 > 首先我想讨论的是"快速制造"现象。如今我们可以用"快速制造"这一标签阐述这些成指数级发展的技术。

尽管快速制造（Rapid Manufacturing）及快速成型（Rapid Prototyping）概念诞生于上世纪九十年代，但近年来才得以飞速发展。因为还处于试验阶段，所以这些制造方式是多样化且彼此不同的。总之，快速制造是一种可以添加原料并通过三维数字化 CAD 模型生成产品的加工工艺。它通过二元平面向机器传递信息，放置原材料并用激光光束对其烧结。快速成型工艺具有良好前景：它不受产品数量约束；也不受产品外形限制（目前来说唯一受限的是产品尺寸）；同时可以进行实验研究以寻求最适合的材料来减少浪费（甚至可能实现循环）。快速制造除了带来了技术上的突破与经济上的腾飞以外，最值得研究的就是制造工艺的本质变化，即对产品（及其消费）进行定义。我们已经说过，自设计学科诞生以来，区分设计师职责的一大特点就是大型工业化生产过程中，设计师能够对生产成果（产品）进行定义，并向其他参与人员传递特定的信息。换句话说，设计师相当于被剥夺操作程序的手工艺人。相反，得益于快速加工技术，设计师重新占据生产领域。他们通过操作产品的三维模型，重新回到类似于工业时代之前的手工作业状态。设计师借助信息工具，缩短了设计与生产之间的距离，直接且快速地获得最终产品。

按需设计 > 开放源代码（open-source）概念诞生于上世纪九十年代后期，并逐步替代了自由软件（free-software）概念，相对于产品（实际上指的是自由软件）它更加注重开放"源头"（信息），让每个人都可以对其进行访问、补充、修改以及完善。以此激活了一个独特的"创造性"过程，在这个过程中形成了无限潜力的共享网络，人们借此实现分享、操作和传播知识。无论是有形的还是无形的（想想CopyLeft和Creative Commons取消了作者版权）产品，都已不再是个人或者独立公司的成果。至今我们所说的"开放式设计"是在新产品开发过程中，面对新产品带来的无止尽的知识，个人不再拥有对它的所有权。设计、生产和消费之间关系的不断演变，涉及的形式与技能（部分或完全）的变化，可以围绕三个关键概念来表达：桌面制造、完全控制以及智能产业链。其中，我认为最有意思的是对设计、生产和消费关注点的转移，不仅仅停留在创新产品阶段，而是更加注重创新流程。我们也知道，创新产品把注意力集中在生产过程的结果上，即关注生产本身，意大利设计在这种情况下，借助创新赋予产品更多的文化价值。然而，创新流程指的是操作过程，它将设计、生产和消费相关的人员与机械置于首位，因此，创新流程是蓬勃发展的。正如我们所看到的，快速制造技术的发展与普及开辟出了一条新的道路，形成了由少数人的独特专业向大众的协同智慧的转变。对于设计和企业而言，这也就意味着设计师和企业之间的合作模式已经过时，他们更应当充分利用网络深入发展。信息技术的传播将重心从构思（单纯的设计行为）向设计与生产结合方向转移，它非常接近于DIY（do it yourself，现在更好的版本是do it with others），其中设计师实际上成为了制造者，并有机会对流程本身进行全面控制，并以这种方式实现持续的实验，而不是等待"客户"；同时，设计师从信息输出者转向输入者；最后，即使在传统行业（如意大利制造），通过引入数字化生产的管理流程使得产业链从解决问题模式转变为灵活智能模式。

将现象简化为单独知识的策略（现在主义源泉），逐渐变为将现象与背景环境相结合（后现代主义），进而实现了范式的转变。所谓的背景环境是指网络世界又或是简单的一个房间 [温伯格（D. Weinberger）的比喻方式]，在这个房间里不需要聪明绝顶的个人，而是要集合大众的智慧，将每个人的思想、辩论和阐述汇集并应用于背景环境当中。因此，设计、生产和消费之间的关系由"单向状态"变为"需求状态"。也就是由原来的一个设计师面对一群客户变为设计师和用户一一对应的交互模式，其中涉及的内容与技能也灵活多变，以不断地适应参与者的具体需求。

Tonino Paris

Francesca Angeletti translate

Design
texts and contexts

Design_texts and contexts
author
Tonino Paris
Francesca Angeletti translate

I would like to express special thanks to Dr. Shaonong Wei. Without his heartfelt participation, I would never be able to complete this book. I am sincerely grateful for his encouragement, constant input on the content and expression style, which play an essential role for the publication of this book. In addition, my special thanks also go to Jingjing Lin. Her patience and dedication to translating this book are highly appreciated.
I would also like to thank everyone with whom I shared, over many years, the intellectual labour that, with the journal DIID_*Disegno industriale|Industrial Design* and so many events to disseminate the culture of design, has generated ideas and reflections, and given life to my way of understanding the discipline of design – which is to say as an instrument of knowledge, a proactive tool to improve the life of human beings and their social relationships.
With so many people to thank, I would like to extend my particular gratitude to Vicenzo Cristallo, Cecilia Cecchini, Federica Dal Falco, Loredana Di Lucchio, Lorenzo Imbesi, Sabrina Lucibello, and Carlo Martino, for many years as my group of fellow teachers. And lastly, I thank the students and teachers at the school of design at Sapienza University of Rome – a school that I founded, and to which I devoted my time, repaid by so much that I received in return.

Foreword

As early as the 20s and 30s of last century, with the development of urban modern commerce, some Chinese schools have teaching subjects related to design, such as Advertisements, Posters, Arts and crafts. But Design, as an independent discipline, became a systematic teaching subject gradually in the middle of the 80s in Chinese universities. It was coincident with the time when I started to be involved in art education. The opportunity to gain deeper observing and understanding of this discipline came more than 10 years ago, when I was occupied as dean of School of Design in university.

Today, there are more than 200,000 students studying design in Chinese universities every year. This is a rather amazing number, but the problem we are faced with is who to teach these students. Learning design is not easy, while teaching design is still a more difficult and challenging task. Obviously, we cannot find so many qualified teachers in a short time, and we even cannot find suitable textbooks for students. Currently, many of teaching materials in use cannot guide students to understand design correctly. On the contrary, they will mislead students seriously, which is a very worrying thing. In fact, texts and contents are significantly important for studying and debating issues related to design.

Six years ago, School of Design from East China Normal University hosted a students' works exhibition in Rome, Italy. In the exhibition, I met Prof. Tonino Paris, director of College of Architect, for the very first time, on the campus of University of Rome, Sapienza. Since then, we collaborate and communicate with College of Architect from University of Rome and Prof. Paris closely on academics and education. Many years ago, Prof. Paris founded and directed *Disegno Industriale | Industrial Design*. In his own words, he hopes he could reconnoiter "true dimension of design" from both historical and contemporary perspectives. DIID, *Disegno Industriale | Industrial Design*, is the journal founded in 2002, which has seen the involvement of more than 450 researchers, scholars and professionals on the Italian and international design landscapes, to investigate, with about 850 essays, the phenomenology of design and the network of its important figures. Each issue has dealt, in monographic form, with a specific subject articulated in the following themes: innovation of the processes of industrial production; the progress of design practices; the study of the development of the languages of the applied arts; technological, scientific, social, and economic conditioning in the construction of the artificial environment; the evolution of the results of research and of theoretical and design experimentation for the development of material and immaterial artefacts; the technology, science, society and economy development jointly in the construction of artificial environment, and etc. The discussion on these representative themes in the design field，represents a layman's debate on the technical and cultural extension of design as a product system. These moving targets were attained with compound results. Prof. Paris used DIID as an operational tool in the field capable of discovering and recounting design as a space for man—in the sense of "extension of man" in order to help readers understand the physical and allegorical implications in design, and determine this "extension". In this way, it shows how design can covert people's needs into products, how it improves the life and relationships of everyone, and figures out the unique meaning of the times of design.

Two years ago, Prof. Paris sent *Design_testi e contesti* as a gift to me, which was just published. I found this book covers a particular area of his academic work, summarizes his thoughts on the themes of the design of the artificial, including the value of innovation, the social dimension, and so on. It also describes the meaning of design, especially when he tries to discover the substance of design and defend it with more deep information from the lexical content. In a certain sense, the book of Prof. Paris is more like a critical rereading of DIID journal, which has been directed by him for many years. Because of this main objective, Prof. Paris converted the central portion of the book into: texts and contexts, which was completed in the means of anthology. He selected a sort of anthology identifying new categories and signalling his interests. This could be related to his work on teaching that he still makes efforts on. This book is like an internet hypertext, but Prof. Paris has thought to unite these clastic or mosaic paragraphs metaphorically within the frame of a single "design discourse" in order to make it more legible. These discourses supported by the introductory essay constitute the first part of the volume, Designing an artefact _ topics. In the volume, he expresses himself in the first person. The last part, Compendium, critically originates from essays or lectures by professors whom he has been working with for years. In the end, Prof. Paris critically discusses the culture of the design in desk's style. At the same time, he points out the methodology about the practice of modern artificial design, and describes the way of comparing and studying theory, method and material by practice.

This book is an important literature work, regardless of the summary of academic career of Prof. Paris, the modern design, and design education community. Prof. Paris attempted to use design discourses to evoke the discussion on design, emotion from design work, thoughts on the relationship between humans and nature, and eternity of design. The critical debates evoked by this book express not only continuous dialectical dialogue about sustainable design but also Prof. Paris and his colleagues' passion for the culture of design, which is delivered to the young generations.

The readers of this book are obvious, including people who are interested in design and its culture, and those who teach and study design in universities. To whom, design is an experimental place, and a world with unlimited developments and possibilities. The logic and rhetoric of expression are the essential components of this book, which play the same role as the contents, and which can easily get lost through translation. Thus, we publish this book in both English and Chinese for complementary reading. Special thanks to Prof. Paris for the Chinese copyright of this book. Great thanks to Mr. Guangye Ruan and his excellent editing team from East China Normal University Press. Many thanks also go to Ph.D student of Prof. Paris, Mrs. Jingjing Lin, who is a junior lecturer majoring in Product Design in School of Design at East China Normal University. It's such an amazing job that she could accomplish the tough work of translating the original book in Italian within half a year.

The publishing of this book can be seen as a gift from a person with passion on design to those who are also enthusiastic about design.

       Prof. Shaonong Wei, Dean of School of Design, East China Normal University
       October, 2018
       In Berlin

Table of contents

Table of contents

p.11   [1] **designing an artefact_topics**

p.29   [2] **texts and contexts**

p.32   design and industrial design
           ubiquity
           design autonomy
           inevitably industrial
           teachings in doing
           matter and materials
           traing and informing
           difference and identity
           the event of the exhibit
           virtual architectural complexes
           it is design
           the language of the object
           prospective scenarios

p.45   conceiving consumer goods
           new technologies
           colour
           teaching methodology
           environmentally sustainable technologies
           connections
           ambiguity of meaning
           training experience
           market demand
           spectacularization
           communication
           deception
           thinking materials
           magnetic cards
           useful obsolescence
           playful appearances

p.57   things, objects and products
           complex paths

without interruption
urging ingenuity
architectural object
strategic value
at the service of sport
anonymous but industrial
between objects and contexts
sincerity and ambiguity
technological innovation
hybrid products
design museums
upgrading

p.66 ethics and aesthetics
the toolbox
doubt
training
the agon of a smart culture
homo faber
performance
into chaos
how, what, who, when, why
light or strong
the reasons
the aesthetics of high tech
coexistence of systems
ethically-minded
the dominant taste

p.78 useful and useless
lifelong education
memory
the crisis of architecture
an integrable design
a new balance
a large farm
interconnections with man

practices
the designing of food
cognitive itineraries
the rare and the exclusive
an octopus

p.89   hybrid realities
lectio
cross-pollination
interaction design
hybridisation processes
design for cinema
social nature
transfers
nanodesign

p.97   innovation through culture
social chromatism
semantic value
design schools
conjugation
productive territories
italian style
italian look

p.106   traditions and transformations
a lesson
units for architecture
new ferments
between two lands
popular design
aestheticising
no-name
the flow of products
cultural complexity

p.119   [3] **compendium**

p.121   naming design
          articulations
          artefacts
          the biography of things
          creativity/verbal fetish
          the ethics of design
          the crisis of design between the modern and the postmodern
          materials in the culture of design

p.132   the word "type" for design
          they are called "types"
          the distinctive traits of objects
          type and modern movement

p.135   design: new paradigms

          design in the open source era
          the design of desktop manufacturing
          design on demand

§
[1]

Designing an artefact_topics

§
[1]
p.13

Designing an artefact_topics

Note > *The book begins with "Designing an artefact _ topics" because it critically debates those issues held together by two red threads that travel in parallel and, when intersecting, complete one another: one holds together the issues typical of design culture, the other the theories, methods, and operative practices inherent to the design of an industrial artefact. This is done subjectively, with arguments in part complete and in part incomplete; with metaphors that evoke the profound spirit of the design action; with reflections on the unvarying subjects of design, and on the variables in relation to the contexts; and with a gaze turned also towards the design of the near future.*

In each case, it is always quite clear for whom the book's reading is intended: for those who are interested in design as a locus of experimentation and a factor of development; for those who, at design schools, teach or are learning how to grapple with designing.

For some time, my chief concerns have included that of not betraying the task assigned to the ethics of teaching, especially when I am discussing design topics. This is all the more the case during these years, years in which my relationship with the school, in all its forms, has become difficult, and often tiresome.

The reasons, the more personal ones, survive in the fragile and fatal rhetorical combination in which the figure of the teacher and the faces of the pupils — together in a classroom not necessarily composed of walls — share useful knowledge. And "useful" is a demanding word, especially when examining the complexity of the task we take on in having to train, through the use of theoretical models and practical activities, those who wish to design the artificial world — starting from an artefact. From here on, as a natural consequence of the role I inevitably feel I am still playing, I have drawn inspiration to put together a sort of critical itinerary-a subjective repertoire starting from the fact that it covers no assigned topic comprehensively, but instead follows a personal, "incomplete method." This is likely the only method that can give balance to certain narrative needs dictated by my variable interests in dealing with certain untarnishable convictions. While what is read may appear, by my choice, fragmented in content, my ideal interlocutor — whoever is interested in design as a locus for experimentation and factor of development for looking towards such a future for all-would not be.

• • •

Context > Knowledge, but above all understanding of the context in which the design action is expressed, is the first and necessary passage for accessing the complexity that the conception of a new work, big or small, demands. However, for a designer, context is not to be immediately and superficially exchanged with that of an environmental nature we are conventionally used to. That one is always there. It is a rhetorical locus — insidious but foundational: the "cultural territory of design." And, if I think of the current one, I cannot help considering its slipperiness. While apparently free of encumbrances, it is rife with deceptions. It is an ambiguous area, because its nature is fleeting. I am referring to the Postmodern, the culture of the temporary. An invasive and interstitial culture that is the mark of our contemporaneity.

Umberto Eco effectively outlines its role and effects:" The Postmodern marked the crisis of the 'great narratives' that they thought they could superimpose a model of order on the world; it instead devoted itself to a revisitation of the past, and in ways intersected with nihilist impulses; the Postmodern has represented a sort of ferry from modernity to a still unnamed present; a present in which an entity that guaranteed individuals the possibility of solving, in uniform fashion, the various problems of our time disappeared. From here on, the crises of the ideologies were outlined, as well as, in general, the crisis of each appeal to a community of values that allowed the individual to feel he was part of something and interpreted his needs. With the crisis of the concept of community, an unchecked individualism emerges, where no one is anyone's travelling companion anymore, but an antagonist to be watched out for. This 'subjectivism' has undermined the bases of modernity and made it fragile: everything is dissolved in a sort of fluidity [...] and the only solution for an individual without reference points is appearing at all costs, appearing as a value, in a sort of bulimia with no purpose other than to scrap the old in order to take part in this orgy. What can replace this liquefaction? We don't know yet, and this interregnum will last a rather long time [...] The trouble is that politics, and to a great degree the intelligentsia, have yet to comprehend the scope of the phenomenon." [1].

This is a reading of contemporaneity that, in order not to remain a *vox clamantis in deserto*, must produce, in each of us, a new and impassioned propulsive energy: it must encourage our renewed attention to the content of the things we design, rather than to the ephemeral dimension of their appearance. In doing design, man must return to being the focus of our objectives. The design of the artificial must be restored the simple and inalienable ability to interpret man's needs, in order to transform them into products destined to improve his life and his social relationships.

In these latest times, I have noted recent and special conditions for a critical space within which to glimpse an almost philosophical support for these aspirations of mine on the function of design, in two evocative events: the 14th Venice International Biennale of Architecture, curated in 2014 by Rem Koolhaas; and the exhibition "Neo preistoria—100 verbi," curated in 2016 by Andrea Branzi and Kenya Hara, on the occasion of the 21st Milan Triennale.

At the Milan show, 100 verbs express 100 human actions: the invariants to which man, in his evolutionary space, has made precise instruments correspond. We are thus moving among inventions and objects that have marked the development of human civilization. The verbs that are used recount the practices of measuring, of navigating, of writing, of cooking, of playing music, of killing; others, on the other hand, express feelings, intangible abstractions, but all the same expressions of invariants that, in the history of mankind, have been formalized in objects-and thus terms like to love, to charm, to imagine, to invent, to think. Life is associated with the aspirations of man, in whose evolution they are expressed with the desire to eat, and thus to cook, with the ambition to fly, with the desire to communicate-which is gradually expressed with signs, or by looking into each other's eyes, and then with words, and later with writing, and nowadays with all those media capable of doing all this without barriers. At the same time, contact with death cannot be avoided, which is why the list also includes such verbs as to annihilate, to pierce, to threaten, which evoke all the evil that can be generated in the man's mind, with no difference between past and present. For both, each activity articulated by verbs is thus made possible by the invention of new objects that are increasingly evolved in performance, form, and technologies. As Branzi puts it, scrolling through 100 verbs in 100 instruments represents "the flow of a vital and mysterious energy that, from the darkness of history and of an infinite space, through the frontiers of scientific research, expands human survival through the production of our body's spare 'parts'" [2]. For Kenya Hara, these "spare parts" give material form to man's desires because "humanity has developed its skills through utensils; and at the same time, through utensils, it has increased its desires, so the 'history of man's desires' may be understood as the 'history of utensils'." [3] And now that we are in the 21st century and are fully proceeding towards the age of artificial intelligence, it is very important to reflect on what evolution humanity will have, and in what situation it will be found, and what verbs and utensils humanity will need, by seeking the reflections of the wisdom and errors of the past.

Branzi and Hara's lesson is one that goes deeper into, explores, and exhibits all the design skill of *homo faber*, and the knowledge with which he has improved his articles of use. But it also leads us to reflect on the need to interpret man's actions, even in the smallest object. While the technological differences separating prehistoric objects from contemporary ones are clear, the same cannot be said as regards their design status. The will to shape matter according to one's needs creates a thread that holds together the entire universal history of the human being. Materials and technical/scientific possibilities change, but the skill and ingenuity to invent tools capable of satisfying desires and meeting always new and unexpected needs do not. These are needs that people in design must work alongside of with the necessary humility in order to configure real solutions, and not the result of the meandering formal self-referentiality of-as we would put it today-the starchitect.

In the very theme of the Biennale, *Fundamentals*, Rem Koolhaas clearly distances himself from the figure of the architect in his extension as starchitect, and returns to devoting attention to an architecture rediscovered starting from its own foundations. He restores these foundations in the manner of an exhibition, recreating a special sensation of a museum of Architectural Histories. With *Elements of Architecture*, in the Central Pavilion, Koolhaas explores the genealogies and mutations of architecture's individual constructive components. He digs down to the foundations of the past in order to investigate the present and be projected with clarity into the future. With this sort of rereading of the alphabet of architecture, he appears to uphold a message of hope in implicitly stating that the culture of modern design has not been entirely emptied of identity and left flattened under the stresses of globalization. Koolhaas is asking us to overcome the effects on the form of the city produced for so long a time by the strong crisis of ideologies, that spoken of by Eco (from Weak Thought to Deconstructionism, from Relativism to the Postmodern). Or perhaps he is asking us to encourage architecture and design to take renewed interest in the subject for which they are ethically destined: man. And for this to take place, Koolhaas believes it is necessary

[1] **Design_texts and contexts**

to take a new look at the constructive invariants of homes, of buildings, of the city — inalienable elements of a material DNA, used by every architect, always and everywhere: floors, walls, ceilings, windows, façades, balconies, hallways, fireplaces, staircases. It is an exercise that is naturally concluded by traversing patterns, models and established typologies, in order to later experiment — starting from the need to offer new models for the new man — with the new technologies to the extent needed.

Type and model > Both material and immaterial elements come down to notions of type or model. The first is the historical consolidation of elements whose constitution, although subject to evolutionary processes in forms, functions, and technologies, is immutable. The second is the formalization of a new arrangement, designed to respond to new functions for man's new needs. The model is therefore the reference from which new types may be born. Both the type and the model are likened to the fundamentals, to the constituent materials of both architecture and of articles of use, and in their turn of the *forma urbis*, in a strange blend of flow and persistence that has endured for more than 5,000 years. The fact that the elements change, one independently of the other, according to cycles and economies and for different reasons, can therefore transform architecture and the article of use into a collage: complex, archaic and modern, unique and standardized. Type and model, if re-examined in their constituent invariants and organizational arrangements, allow us to recognize cultural preferences, forgotten symbolisms, technological developments, political calculations, regulatory requirements, and so much more.

The translation of the notion of type as constituent material in the sphere of design as well does not come immediately. It is a translation that, for articles of use, passes through the verbs evoked by Andrea Branzi and Kenya Hara: they are actions, activities, human desires — that consolidation of which historical invariants are made. It is no accident that articles of use respond to man's needs as if they were prostheses for his limbs and minds. This is what took place, for example, as man defended himself and went on the offensive, availing himself first of elementary rocks and gradually going on to more complex objects.

As may be seen, it is all a succession of artefacts, of articles of use, that are renewed by making the previous ones obsolete starting from their formal and classificatory nature. In fact, "both in the phase of their design/fabrication, and in that of their recognition/use, part of our knowledge of artefacts is a 'typological knowledge.'" Without this permanent framework of things, we would feel dispersed in a universe of unrecognizable material substances we would not know what to do with. We need every design to have and to maintain, to the extent possible, its own identity, or "that particular formal, structural, and functional configuration that sets it apart from the other technical objects [...] Design, for its part, determines the distinctive characteristics of objects, and builds their identity through the specific configuration of each technical element. Their configuration, then, makes them recognizable in any variant they have within a typology of objects, especially at times of strong technological innovation [...] "[4]. However, this taxonomy shows its inevitable incompleteness in the face of what has taken place in the world of research, of production, and of products in recent years.

• • •

Téchne and artefact > Of all animals on Earth, man is the one that, by his physical conformation, by his sensory apparatus, is the weakest, and thus destined to succumb in the daily struggle for survival were he to be forced to grapple with any large predators in nature, with his bare hands. If we consider sight, for example, we find there is no comparison between the visual acuity of certain animals and that of man, and even less so if we measure his ability to adapt to darkness.
Some beasts see at distances ten times, a hundred times greater than those man can, but above all they have the exclusive ability to appreciate, through sight, both the path to take and the amount of time needed to reach a designated point. This ability is an extraordinary tool in hunting. It allows them to assess, for prey they have sighted, the time needed to approach it, and thus whether or not it can be captured, given the distance from which the chase begins. And yet it is man-who is not barehanded-who prevails over animals, because he uses his ingenuity. And he uses it to conceive and develop prostheses of every kind, whenever it is necessary to make up for the most disparate shortcomings. Starting with tools, from clubs on, conceived for initial, basic defence, human beings have gradually produced instruments increasingly useful for the purpose: the axe, the knife, the arrow, and so on, down to the revolver and all that followed. Therefore, the distances man does not cover with his sight he covers with glasses or binoculars, or with apparently invisible devices like contact lenses with which he can alter the colour of the iris, thus modifying the aesthetic nature of the eye. A prosthesis is thus used to introduce apparently superfluous needs, if we consider the primary ones they probably stemmed from.
This initial discussion of sight and prosthesis introduces — and, it would appear, simplifies — the concepts of téchne and artefact. Téchne, in our case, is the use of instruments and procedures suitable for obtaining a product — a prosthesis, we might say — to improve the conditions for remaining in the environment we live in. This process of transforming a need into a prosthesis is the "simple" result of man's ingenuity and artifice. In particular, with ingenuity, by defining its performance requirements upstream, he designs a product that responds to a declared need; on the other hand, with téchne, he prepares and controls the materials suited to the pre-established purpose, the instruments, and the working methods adopted for making the artefact. When the sequence just described dissolves into the direct participation in the act of execution, the separations between those who theorize and those who do are cancelled. And it is the vision typical of the so-called artisinal product, which is to say the result of a process in which "the activity of thought" corresponds to a heritage of cultural references and knowledge related to circumstances of time and place, while the "activity of doing" corresponds to the employment of methods and instruments suitable from time to time for transforming materials into objects, by means of aptitude and ability acquired through training and apprenticeship.

[1] **Design_texts and contexts**

The work has no autonomous artistic validity, but at most expresses a collective aesthetic intentionality. It is a practically unlimited path: the product's requirements are not defined once and for all; the artefact is therefore always susceptible to further improvements, also by subsequent generations. Téchne is then developed by gradual accumulation of experience, on the basis of an ongoing tradition. It is easy to take the products of the material culture and of the applied arts as an example of this concept. For instance, the domestic tableware that maintains constant the function of collecting, preserving, and cooking food products. These are artefacts whose improvement takes place slowly, over generations.
In this regard, I would like to discuss some children's toys as emblematic of a collective aesthetic and technical knowledge. In an impoverished region of our continent, I have seen a toy similar to a yoyo, made with a stick, shrubs, a rope, and a piece of a fruit rind, called merely "*briquedo*" — that is to say a general definition to indicate the object as a toy. It was a simple yet complex artefact, bringing together: knowledge of the statics that held the various parts under stress together while playing; the notion of transformation of work into kinetic energy; the basics of the principles of acceleration and inertia; a profound knowledge of the breaking limits of the materials it was made of; and, lastly, the general principles of ergonomics, as seen in the proportions between the grip and the hand holding the toy. This knowledge was nothing more than the gradual "piling on" of experiences passed from one generation to the next, settling in layers upon this object. The child who made it perhaps did not even know how to read. But it is certain that the artefact had deposited onto itself a large quantity of technical knowledge that had become collective heritage.
This also took place with reduced-scale reproductions of bicycles or of other objects that represented an opportunity for play. The model of reference was substituted by a simulacrum that was increasingly refined, increasingly close to the model, increasingly accessory-rich (also at the risk of altering the proportions between the parts), and increasingly varied in colour. The artefact born as a personal game is transformed into a sort of commercial product, the result of a craftsmanship that is paradoxically highly evolved since it can be reproduced unaided by equipment and systems, but by hand alone, and by personal skill alone.
Lastly, there is the simplest of the toys, a truck, an imitation that allowed children to transform themselves into users of an impossible dream: owning and driving a tractor-trailer. And who knows whether the most entertaining game was fabricating it, rather than that using it as such. Here, the artefact let itself be admired starting above all from the material used to make it — wood. "Wood, ideal for its solidity and tenderness, for the natural warmth of its contact [...] , eliminates the cutting of excessively sharp corners [...] , and when the child handles it and strikes it, it neither vibrates nor screeches. It makes a sound both muffled and clear at the same time; it is a familiar and poetic substance that leaves the child in a continuity of contact with the tree, the board [...] Wood does not cut, nor does it go bad. It does not break or wear; it can last a long time, live with the child, and, little by little, modify the relationships between the object and the hand; if it dies, it does so by reducing, not swelling like those mechanical toys that disappear under the hernia of a

broken spring. Wood makes objects essential — objects for always."[5].

It may be maintained that when the conceptual plan and the executive plan are integrated, artefacts are reproduced that are emblematic in technical content and constructive practices, to be transferred in terms of widespread experience accessible to everyone. If the conceptual plan is maintained separate from the executive one, then there is a design and a designer. The use of procedures and tools to make a product then becomes a reproductive or mechanical practice; the operative autonomy of the labour is practically nil: the product is concluded in the design. The design defines, entirely and a priori, the functional and formal characteristics of the artefact, and the subsequent execution procedure introduces no aesthetic value that is not already determined with the design. These include the material execution, which is to say the assessment of the intrinsic possibilities of the technology of the time for the purpose of making the product.

And this is what took place in the transition from preindustrial to industrial civilization, in assessing the production of the daily or technical article of use. And it is even more markedly taking place in post-industrial civilization and the civilization of communication, where it is thought that it is man's instruments and not his ideas that determines them. Right from the first acts of design, the designer's role is decisive in interpreting the results of research and of scientific discoveries, so as to borrow them in technological choices and execution processes, in the better organization of the artefact's performance requirements in order to best respond to the demand that generated them-a demand that does not end with the demand for functional performance.

In fact, there is increasing attention to the product's communicative values, to its aesthetic content, to its symbolic value, to the need for exhibition. Also to the superfluous. To all those values that shorten a product's consumption time, exponentially accelerating the obsolescence of its content.

The prostheses born to survive among other animals and in hostile nature with all its hazards, born as functional apparatus, were then transformed into artefacts that were high-tech, highly communicative, and increasingly intended to satisfy the superfluous.

To remain among the examples that refer to technical objects and the articles of use that have always been the result of the designer's decision-making process, let us consider what shows the time: the clock. The measurement of time is certainly a function that man was soon to find essential in order to dominate the distance between day and night, the cycles of the phases of the moon, the passage of the seasons, the duration of pregnancy, his very stay on Earth, and the social economy. For these purposes, he conceived and built devices that relied on shadows — sundials — or the reading of a material that changed position in a determined period, which is to say the hourglass. Gradually, he perfected the technical content, applying the principles of mechanics in the wind-up watch or the pendulum clock, until achieving absolute accuracy. For a very long time, the parameter of assessment for choosing a timepiece was its accuracy, and it was of little importance whether it was designed by an unknown maker, and whether its appearance held little content.

Then came electronics, an exact science par excellence, and printed circuits-miniscule, extremely low-cost components, onto which an electronic device's operating process is printed by means of industrial processes. With their aid, the technical content of the timepiece became unsurpassable, with an accuracy leaving no room for improvement, and above all at low cost. Something else came into play to define the object's value: its figurative qualities, its symbolic meaning. Its design.

The Swatch was emblematic of this process. Its "original design" was the true factor making it a market powerhouse. It almost immediately became a collector's item, chosen for its design, with no one disputing its accuracy and technical content. And the fact that the particular nature of its design makes reading the time complicated in some models becomes yet another mark of its originality. It is the dimension of the superfluous prevailing over that of the useful. In fact, the object's form is increasingly denied a relationship with the function of use; often, neither the form nor the materials have an organic relationship with the item's intended use.

Valid for this purpose are all the examples in which it is clear that ambiguity and deception prevail in the item's design, and the use for which the item is intended is almost hidden, giving rise to forms of ambiguity fully benefitting advertising. So an octopus is concealing a juicer; vaguely metaphysical shapes emit a light that looks like a lamp to us; we strain to recognize we are holding flatware, rather than artworks, in our hands; forms that deceptively allude to a feather are instead humble toothbrushes. Then there are buildings conceived as objects, and objects made as if they were buildings.

One perceives the clear, strong contradiction that Pier Paolo Pasolini — one of the most important interpreters, and not only in literature, of the transformations produced in the extraordinary transition from agricultural civilization first to industrial and then to post-industrial civilization — was already warning against forty years ago. Indeed, he grasped how the transformation processes were being increasingly agitated within the modern-day contradiction of the terms "development" and "progress"; between ideal political demands and emerging social models. An abstract concept like "progress" — an ethically indispensable factor like the evolution of scientific knowledge and of technologies for improving humankind's living conditions — is placed in comparison with the real dynamics of "development," which is to say between the pragmatic demand and the day-to-day relationship with economic facts, with the production and consumption of superfluous goods. It is a contradiction that is reproduced in the "consciousness" that sustains the ideology of human rights and with it the idea of progress, and in "existence" founded upon the consumerist ideology, and thus upon the values of development.

• • •

Design between state of the art and invention > Continuous verification of the relationship between theoretical and practical thought is central to design experimentation. This means that theoretical thought takes shape only in the presence of practical thought, without which design does not take substance in the finished organism. Theoretical thought arises from our culture, from our set of knowledge, from the storehouse of our memory. Practical thought, on the other hand, is the result of training, of practice, of a continuous verification and updating of technical knowledge through the instrumental apparatus accompanying our craft on a daily basis.

Design fixes, once and for all, the principle of the form and the performance content of a manufactured article rather than an artefact. But design is under the constant surveillance of technique, because it is technique that makes form tangible, whatever its nature is. But without the action of practical thought, or, if you will, of the art of doing, design is a tenuous subject.

Before design, however, comes the plan: for example, the word "vestibule" is written in the plan, and the architect must transform it into a place for gaining entry.

Between plan and design, there is an intermediate space that is consequently transformed into nothing other than the recipe of required ingredients. This change of state starts one instant before the designing begins, but is perhaps already within the design, and certainly represents its foundational moment. It is the moment when the designer — starting from the ambition to govern the relationships between the irrational of the design and the rational of techniques — is alone with his white piece of paper, with the list of ingredients at his disposal, with his storehouse of memories, with his own knowledge of history, of natural and artificial materials, of production systems and technologies, of the contradictions of reality, but with the certainty dictated by three truths: the design's theme, his practical thought, the places and the man for whom it is intended.

Starting from the theme and the place, the generative conflict is developed, which will then continue throughout the creative process. It is a combined conflict of forms, institutions, people, materials, and construction systems, a conflict that finds its resolution in the gradual process of knowledge, through logical construction, always the same as itself: an emotional nucleus that escapes analysis but is connected with the theme and the place, and will grow as the knowledge of the theme and the place grows. This initial emotional datum is associated with a specific typological solution, which is taken apart, analyzed in its functional, constructive, and figurative characteristics, and lastly put back together.

It is the phase in which long experience and in-depth knowledge of the rules of the art of building carries great weight. It is a procedure that aims to give the right tension to the design practice, to the space of exploration that is opened, restoring meaning to artefacts, to architecture, to the figure of the technique that operates in this dimension.

Vitruvius's account of the formation of the Corinthian capital is a complete metaphor for explaining the relationships that exist between state of the art and invention.

Here is how Vitruvius tells the story:

" [...] *A Corinthian virgin, of marriageable age, fell a victim to a violent disorder. After her interment, her nurse, collecting in a basket those articles to which she had shown a partiality when alive, carried them to her tomb, and placed a tile on the basket for the longer preservation of its contents.*

The basket was accidentally placed on the root of an acanthus plant, which, pressed by the weight, shot forth, towards spring, its stems and large foliage, and in the course of its growth reached the angles of the tile, and thus formed volutes at the extremities.

§

[1] Design_texts and contexts

Callimachus, who, for his great ingenuity and taste was called by the Athenians Κατατηξιτεχνος (Catatexitechnos), happening at this time to pass by the tomb, observed the basket, and the delicacy of the foliage which surrounded it. Pleased with the form and novelty of the combination, he constructed from the hint thus afforded, columns of this species in the country about Corinth, and arranged its proportions, determining their proper measures by perfect rules [...] "
Marcus Vitruvius Pollio, De architectura, Book IV
I like to point out that Callimachus (or, "he who exalts material with technique") is credited with the invention of the Corinthian capital in accordance with a combined procedure in which theoretical thought exercises control over proportioning, and practical thought exercises control over the way of working the material, to arrive at the formation of the architectural figure as an act of balance and synthesis.

• • •

Sincerity and ambiguity of materials > The technology on which the fabrication of buildings and of articles of use is based has radically changed over the last fifty years. The building has progressively dematerialized and at the same time been enriched with functional performance features through mechanization, electrification, and computerization. These innovations have brought the effect of separating the construction shell from the bearing structure, thus creating a sort of skin around the building skeleton; they have progressively transformed the unitary and homogeneous nature of the architectural organism, in the building's organization in systems and components: from that of foundation to that of systems; from the structural frame to the envelope. Each of these systems gradually grew more and more independent of one another, each with its own criteria and specialist skills. This tendency is easy to imagine growing more and more with the rise of telematics, an area indicative of the importance given to the efficiency of the systems of comfort and communication, in relation to the physical durability and the representative value of the built form.

One example known to many is provided by the intensive use of air conditioning, even in dry climates, where protection from the sun was often obtained with thick walls, eaves, and cross ventilation, or by building sunshades, and by a general ability to open and close windows and shutters at the right time. Since, in a perfectly sealed air-conditioned environment, it is impossible to open at least one window when the weather is mild, we are disarmed. But the justification may be found in the perverse mechanism by which one technology must provide remedy for a non-spasmodic use of another technology. It may then be said that the application of air conditioning in sealed structures is justified by the pollution arising from the generalized use of the automobile. A perverse, inescapable circle is thus closed.

What occurred in buildings also took place in more technologically advanced technical objects, in an even clearer way: increasingly light, increasingly high-performance, in which weight and performance are more and more inversely proportional. One need merely consider the iPhone, or the MacBook, defined in Apple's advertising as "lighter than light." Apple explains that with the MacBook "we set out to do the impossible: engineer a full-size experience into the lightest and most compact Mac notebook ever. That meant reimagining every element to make it not only lighter and thinner but also better. The result is more than just a new notebook. It's the future of the notebook."

Apple's enigmatic words refer to another contradiction involving the use of materials, in relation to the language they express. With the processes of dematerialization of architecture and of articles of use, alongside the traditional way of interpreting the communicative content of materials, a more critical and unsettling — but, I believe, more effective — way of observing the changes taking place survives. It is the theme I like to define as the "sincerity and ambiguity of materials," and it is a way to affirm that the relationship with materials constitutes, in design, the reference of an emotional tension and of a relationship with the historical memory we constantly have to come to terms with. Materials are the literal constituents of artefacts, as they make them tangible.

It is materials, the entities that have concretely formed human places, that have interacted over time with construction and production techniques, modelling them and modifying them in order to continuously adapt them to their own prerogatives. Materials are the premise of art, and thus powerfully immediate and knowing in evoking a culture in its continuity.

Stone, clay, wood: they are chosen as elements to make "things" depending not only on local natural resources, but on their constructive virtues, their endowments of static power, durability, workability and collaboration with other materials, malleability for shaping, and beauty. Materials are key elements par excellence, live realities that can return to grappling with the technology of our time to be once again reinterpreted and re-proposed in new versions and associations, but also to cross-breed with technology, orienting its experimentation and productive choices through the great strength of their cultural imprinting.

In both natural and artificial scenarios, matter is ambiguous or sincere, vague or authentic. It is ambiguous like the earth or the sand when, aside from the colour, we see nothing; it is sincere when the sand itself tells a story and directs our imagination in a precise direction, or tells us of its formation, of events, of the nature or of the person that gave it shape and colour, the profile of a woman, the characteristics of sphinxes. Matter is sincere like the rippling water announcing to us its undulating motion driven by currents; or it is vague as in the abstract drawing that the water itself paints in a fresco without time or place; it is ambiguous when, as in a sheet of metal attacked by rust, it dematerializes, lacerated by corrosion and dotted with the shapes reflected by the play of lights and shadows; or it is sincere in the spare, undulating metal sheet, where the form speaks to us of a stiffening system that gives it a static efficiency allowing it to stand vertically. It is ambiguous when, with iron, it evokes geometric outlines that are fascinating in their design but do not communicate performance and functions, or it is sincere when, with the spiralling geometries of springs, it evokes univocal functions. It is sincere as in the iron of gratings that, with their grid pattern, trace in space the account

of a trilithic system that is transformed into a three-dimensional arrangement; or as in the wood of a structure where the history of all the working operations it has undergone for its construction is recounted by the weaving of the geometries and outlines that constitute its shape. But it is also the Earth, from which everything originates, that opens space for flights of the imagination: journeys through landscapes that the cracks of a quick dehydration have made similar to the deep furrows of canyons, or natural highlands burnt by the sun.
It is the walls of compact limestone quarries that, in the few traces of processing, tell the story of mankind's arduous labour to transform nature into artifice: squared blocks of stone placed on site by geometric rules, capable of transfiguring a natural depression into a large-scale architectural event, like the theatre in Syracuse; or to transform, without interruption, a rocky cliff into the base of the smooth wall of a human settlement.
More generally, it is a matter of converting the quarried matter into walls, vaulted systems, and articulated spaces in which the material conserves this dual vocation: a rough, sincere character when it is used in forming pillars where individual pieces of stone are counted; sincere again when the material designs capitals composed of stones sculpted and juxtaposed in apparent equilibrium; and sincere yet again, as in Gaudi's Park Guell, where artifice appears illusory and nature shows itself as the true maker of space.
Material is sincere yet again when it tells us the story of the single tower of a pillar at the Temple in Paestum, with its grooved section, the result of skilled, refined chisel work, which gives consistency to the columns by capturing light through chiaroscuro, or in the more elementary construction system: two columns and a beam transfigure the trilith — the most elementary of construction systems — into an artwork, into an archetype. Material is, on the other hand, ambiguous when the pillars become geometric columns that, behind the gleam of the surfaces, conceal the account of the albeit difficult working operations of the maritime theatre at Hadrian's Villa. The material tells a sincere story when, in the very procedures of its fabrication, a productive design is required, a modular and serial principle underlying, for example, the invention of the minimum unit of masonry: the brickwork that is split into two areas: rational production that requires time; and the flexibility permitted by the weight and by the small dimension of the piece, which determine the regularity and the intrinsic rules of brick construction, so as to make a brick wall the form of its work site. Contrary to what is evoked by the brick wall, or by the stone material through which, beyond the size of the pieces, one may glimpse all the working operations that gradually determined its final form, in the technologically advanced curtain wall matter is profoundly denaturalized. Eaves disappear, the metallic weaves become as thin and gleaming as possible; the abolition of all architectural intermediaries with their changing sections and their "gradations" of shape, material, and colour, transform the façade into a single, glass surface with the building's corners rectified and the windows flush with the wall.
In a sort of leap of scale, in which the elementary piece — the glass pane — becomes the reference for the final outcome, the building is proposed as a paradigmatic piece of a technology that, in its

refinement, has the ability to represent an articulated apparatus of elements that contain or conceal the bearing structure of the building, which holds a complex performance apparatus. And for objects, via a sort of transitive property, the same applies. For them, too, the sincerity and ambiguity of the materials determines factors of ambiguity and excellence.

• • •

On the design of the near future > How will the way of designing change in the near future? Clearly, in practice, it has already for some time shown a strong discontinuity with the past, in the ways of designing and producing industrial artefacts. It is a phenomenology that may be ascribed to what, with regard to the productive environment, we define as post-Fordism and that in the cultural setting is placed within Postmodernism. It is a complex pairing that leads, through an inevitable technological itinerary, to open-source civilization. This phenomenon is produced by new information sciences, to which 3D printers bear conscious witness: objects, not even particularly complex, able to transform an idea into physical form, in real time. In this way, the classic sequence attributed to the physical conformation of products is interrupted; it is a sequence marked by the need to involve a variety of professional figures in the vertical supply chain that, from design, continues through production and ends in marketing. The supply chain, when it exists, is now horizontal, and is metaphorically unbound from the need for large-scale mass production, given its tendency for productions that are potentially customized and in the desired quantities. But beware:" [...] the task of design in this case must be more ambitious. Perhaps it will have to change the way of designing; perhaps its placement in the supply chain of the division of labour will be different [...] ; but the designer [in this case, ed.] can rethink and invent typologies. If on the other hand the perspective is that each of us can, with a home printer, become a designer and design our own objects, the risk might be not that of an endless explosion of variables, but of a radicalization of typologies. In the end, this would lead to a stereotyping of our material culture [...] " [6].

What I perceive is a wind that moves in an obstinate direction and counter to the demands for renewal, which is to say to the need to rid ourselves of the harmful effects of Postmodern culture, a synonym for a liquid society so greatly held up as a useful model for enabling new models of participatory democracies, but that instead made it possible, due to the lack of cultural constraints, to replace "big narratives" with "small" ones.

The aforementioned "information society," "post-Fordism," and "Postmodernism" foreshadow beneficial effects in society, due to the spread of a presumed democratic well-being. However, their fatal and combined synthesis appears to me to betray their theoretical mandate and, conversely, shows us a unique and updated extension of the Taylorization of labour and society, with the tendency to set living time and working time in conflict with one another, in a progressive social confusion.

Indeed, a dual and contradictory reading is possible: believing in the spread of small activities rooted in the territory and still marked by a human relationship between employers and workers, or in large industry trying to break free of labour market regulations and trade-union controls. But can the protectionism of a new localism truly slow a thirty-year trend and induce multinationals to repatriate factories, to relocate on national territories jobs that had ended up in China, Mexico, or India? Apple, the Queen of hi-tech, has a productive and logistics chain based on calculations of cost, and also quality. It assembles in mainland China, but integrates sophisticated components manufactured in Japan, Taiwan, and Germany. Bringing that galaxy of branches and suppliers back to a single place in a single country would be a long and costly operation. This is unlikely.

Capitalism, then and now, must grapple with a radically different demand, marked by a fragmented and varied population of consumers. Arising from this is the need for flexible specialization, which is to say the ability of producers-aided by new information technologies and by adaptable labour models-to modify their output depending on the change in demand, even if from small and different groups of consumers. This condition, which favours small enterprises (but not only), has produced, according to post-Fordism theorists, a gradual "disorganization" of capital, which is impacting the production of goods in addition to the organization of social relationships.

In this unstable landscape, the technologies and devices supporting design will change. In support of things, the things themselves may respond differently to new needs we can only imagine today. But there is one need I hope will not change: for the designer to grapple with that emotional nucleus-which precedes the technical one and complements the aesthetic one-that gives rise to the ethical reasons of a design act resulting from the moral responsibility the designer bears, and that leads, in the best-case scenario, from the concept to the work. I believe that even the most advanced technologies will bear real fruit only if they are useful instruments for the quick evolution of ideas destined for humankind.

[1] Umberto Eco, *La società liquida*, in:"Pape Satàn Aleppe"; La nave di Teseo, Milan 2016.

[2] Andrea Branzi, *Neo-prehistory: 100 verbi*, in:"Neo-prehistory: 100 verbi," Triennale di Milano and Lars Muller Publishers, Milano 2016.

[3] Kenya Hara, *Poesia sul desiderio degli essere umano*, in:"Neo-prehistory: 100 verbi," Op. cit.

[4] Raimonda Riccini, *Il senso del design per il tipo*, in:"Type & Model | idee, progetti, azioni," Planning design technology Journal no. 4, Rdesignpress, 2015.

[5] Roland Barthes, *Giocattoli*, in: Miti d'oggi, trad. Lidia Lonzi, Einaudi, Torino 1975.

[6] Raimonda Riccini, *Il senso del design per il tipo*, Op. cit.

§
[2]

Texts and contexts

Introduction

> design and industrial design
> conceiving consumer goods
> things, objects and products
> ethics and aesthetics
> useful and useless
> hybrid realities
> innovation through culture
> traditions and transformations

In these latest years, if I look at the experiences I have had in exercising the theoretical dimension of design, I recognize that some of them have been accompanied by the method of anthology: the collection of essays, or of selected pages, of a work, a designer, a writer. I discovered its intrinsic scientific utility alongside its educational one, and above all the sense of responsibility to be borne by those who work anthologically in the texts of others in order to seek an intellectual balance between the author's words and the sense of individual selection. In the case in point, appearances are deceptive and, even if those who were the object of investigation coincided with those who carried them out, there was no shortage of uncertainties, which were actually duplicated given the obvious difficulty of acting with the necessary critical distance. The result, already illustrated elsewhere, was to proceed by using new interpretation keys capable of reconnecting the excerpted passages within the frame of the word "teaching." The method used was that of taking apart and sewing back together what I chose to re-examine, and I hope this mode of storytelling will persuasively restore the value of chronological summary along with that of a personal and still responsive testimony of my design.

Design and industrial design

ubiquity
from Il design è ubiquo | Lectures 1, 2013

design autonomy
from Dispute vere e presunte, DIID 57|2012

inevitably industrial
from Naturalezza industriale, DIID 55 |2012

teachings in doing
from Design fra identità e diversità, DIID 42–43|2009

matter and materials
from Materiali e natura, DIID 38|2008

training and informing
from Designer After School, DIID 32|2007

difference and identity
from Difference e Design, DIID 24–25|2006

the event of the exhibit
from Fairs for Design, DIID 17|2005

virtual architectural complexes
from Frontiere della grafica e della comunicazione visiva e multimediale, DIID 16|2005

it is design
from Design is everywhere, DIID 7|2003

the language of the object
from Global design/Ethics plus, DIID 3–4|2003

prospective scenarios
from Il disegno industriale, DIID 1|2002

A cautious dichotomy — superfluous and advantageous as the case may be — accompanies the use of the expressions "design" and "industrial design." The former is more flexible, dynamic and outside of time; the latter is clear, univocal, and historicized. While industrial design represented the emancipation of design models in order to conceive the products of modernity, design has become a system for representing, even before the individual products, the needs that generate them and the services that qualify them. It is not a matter of choosing between the two positions, but of grasping the differences in how a design's story is told, while being fully aware that this starts from the choice of words.

Design and industrial design

ubiquity > [...] Design is ubiquitous because by changing its appearance it tends to be a *Zelig*. Design is ubiquitous because, in overcoming all postindustrial forecasts, every day it builds new things and new forms of expression in things. Design is ubiquitous in embodying ethical and salvific positions, formalist drifts and drifts of product categories, all at the same time.
Design is ubiquitous for how it is shaped with respect to the demands arising from the practical culture of industrial design and from the ideological and traditional culture of craftsmanship.
Design is ubiquitous for how it has traversed the economy and culture of industrialism and modernity, of the Postmodern and of post-industrialism, and in fact the very culture of the intangible in the information society.
Design is ubiquitous if defined, according to the interpretation that has been consolidated in the Roman school, as a design of the artificial, of all that is artificial — and then also as representation of the material culture and interpreter of man's inexhaustible needs, as he transforms these needs into products destined to improve his life and his social relations.
Saying that design is ubiquitous — if we consider articles of use in their numbers and the roles they are actually assigned — may appear to be a totally obvious statement. But as a declaration it is only apparently predictable, because the spread and unbelievable metamorphosis of the object landscape that surrounds us is, above all, theoretical. What does this mean? Design has become a global culture which, in the name of innovation articulated in progress, travels the planet without necessarily producing stability, but rather "dynamic equilibria" between economies, societies, and cultures. It therefore does not solve problems, but actually continuously raises new ones. This is not so much a mischievous speculation on the ambiguous concept of the elusiveness of the value of "consumer goods," as the fact that design acts so that the system will never stop, creating the conditions for emancipating the culture of design even before that of the object. Only in this way can it produce research and give itself, beyond any academic and scientific dispute, the sense of a natural need in training.
In the case of training, the ubiquity of design is represented by an incontrovertible fact: in recent times, this is the discipline that has followed, and continues to follow, with greater tension and reflection, the change of professions in the field of designing the artificial, accompanies them, and offers different solutions, accepting the difficult task of grappling with the evolving figure of the designer. No easy assignment, it is the harbinger of uncertainties.
Design is ubiquitous, then, not because objects are everywhere. It is ubiquitous for the quantity of designs that preceded that multitude of objects. It is Quaroni's theory of diffuse design that makes many able to anticipate small changes and modifications, by acting as an aware designer in one's own real world. It consequently promotes change through an interstitial innovation, one without revolutions, without proclamations, but by defending the principle of the possibility of doing-anywhere and no matter what. Objects, then, do not prove theses. If anything, they interpret them, refute them, verify them. But beware: however much this might appear inevitable and democratic, because it is within everyone's reach, the populism of design-of which design is, for its share, guilty-

lies in ambush [...] .
The ubiquity of design promotes various slogans, starting from "design by all, to design for all"-an expression that marks its level of disciplinary and cultural expansion [...] .
Design is ubiquitous because it has "spontaneously changed": it is no longer as it was understood between the 1970s and 1980s, a formal design, a design to solve technical questions by combining them with consistent aesthetic solutions. In the current logics that feed economic transformations, design is if anything the great feeder of the energy of innovation, which is to say "of the great demand to which all world industry refers." With these premises, design and its complex system are not called upon, as Andrea Branzi says, to produce masterworks, but to change scenarios, to create shifts in meaning and then to serve as "permanent producers of offers of ideas." [...] .Design is ubiquitous when it explores the highest levels of technological innovation, from which are born the strategies destined to develop scenarios for the future by elaborating technical objects, articles of use not yet in production, while anticipating new and unimaginable performance by the artefact, new materials, and their unsuspected applications for a market that unites the demand of advanced industrial societies with that of developing societies [...] .

design autonomy > Starting already from the Renaissance (but perhaps earlier), the distinction between designing and fabricating a work, and therefore the roles and functions of the implied subjects, was being emancipated. Today, like then, many artisans have the "ability to do" but might not possess the ingenuity needed to conceive and fabricate something that was not there before. They know how, when needed, to update a consolidated type, but not to replace it. With the arrival of scientific discoveries, the somewhat Darwinian evolution of models (club, rock, spear, arrow, crossbow, revolver, rifle, machine gun, and so on) gradually separated those who design new things from those who make them. In the transition to the age of the Industrial Revolution, this process was completed, involving large-scale numbers in production. At this point, the figure of the designer became complete, and design was stabilized as an autonomous discipline. Walter Gropius, in conceiving the Bauhaus School, must also be connected to these conclusions; although giving rise to an unrepeatable season of transversality between art, design, and architect, he believed in and acted in accordance with the separation of the roles-a division by no accident enshrined by a manifesto written by designers. Mies Van der Rohe, Paul Klee, Kandiskij, and Hannes Meier were a crucial reference for entire generations in the 20th-century debate between technology and the culture of design, giving rise to an endless catalogue of new products, still sought after today by the market, with the acceptance of John Ruskin and William Morris's Arts and Crafts movement. What school, then, can imagine not teaching the design's autonomy from its becoming a product? Especially in the design of hybrid objects, self-design and self own production are certainly contemplated, but we like to consider this cautiously credible (pointed out because they are known), as in the case of Michele De Lucchi who, in his famous "Private Collection," makes this a specific occasion for artisinally inspired para-industrial experimentation.

Design and industrial design

This means discovering and using handicrafts as the rhetoric of making — by committing body and mind — tradition and innovation, " simultaneously," through a process of "osmosis."

inevitably industrial > From buttons to magnetic stripe cards, 'things' that take form as archetypes of themselves offer a natural path to design and a civil, basic solution in which acumen, utility and beauty serve as reciprocal causes and effects. In its critical conduct, it is a phenomenon of industrial design and production processes which proves that no impact is made on authentic good design by the obsession with emphasizing the intellectual property and ownership of creations.
Unsurprisingly, for some time now the topic at hand has been behind a rethink of the sense of creative parenthood of the 'things' conceived and made in a post-industrial system which gives widespread, shared design a metaphorical role to describe the complexity of open-source design phenomena. Makers, self-production and social design are some of the many terms used to label the collaborative nature of modern design, especially when it is necessary to pick up on the cultural and political value that bridges the gap between 'impersonal brilliance' and the new paradigms of utility and beauty which are meant to be accessible to everyone nowadays, regardless of their profile or status.
These thoughts lead us 40 years back in time to 1972, when Bruno Munari suggested giving the 'Compasso d'oro' design award to 'persons unknown'. He claimed that it should be presented to objects with a clear balance between materials, technology and function that made them timeless. They were salvaged from superfluousness because they could be traced back to economy of thought which went from the design phase to the warehouse. If there was any debate about the aesthetics of forms, the solution would lie in the inherent thinking of 'design issues'. According to Munari, the nature of the objects meant that they had nothing to fear from technological innovation and the changing language of design. Indeed, new materials and technology would perhaps allow them to reinvent themselves and endure. They also owe their staying power to the fact that they do not present ambiguous symbols and — most importantly — they do not belong to social classes. They support the possibility and feasibility of the democracy of common sense and the sovereignty of good taste. Looking back, it is clear that this pairing gave an inevitable industrial character to every version of these spontaneous forms throughout their history.

teachings in doing > So, Design (in the sense of industrial Design) and Designers (as the creators of the flow of products that pass through our daily lives) reflect an aspect of creativity that presupposes study in order to know, understand and be able to do: School. So, to understand how creativity is expressed in the Old World, we need to define the identifying traits of Design in Europe, looking at the proposals of our youth, just out of school, who are taking their first steps in their professional life: *Design After School*. We can take a general look at the unique traits of the schools and historic experiences that have been the reference points of so many identifying traits of European Design throughout the past century: from the Scandinavian School to the Anglo-Saxon School, from the Middle European School to the Mediterranean School(s) (Italy, France, Spain and Portugal).

In the Anglo-Saxon School, the identifying traits have roots in the major transformations straddling the 19th and 20th centuries, from the birth of Design with the industrial revolution, a revolution that generated movements like *Arts & Crafts* in England. In alternating positions that oppose industrial and artisanal production, for aesthetic or ideological reasons, which criticise the alienating aspect of work in industry, contrasting it with the creativity of artist-craftsmen, the need to train professionals for the requirements of new times was understood right from the time of Sir Henry Cole. Schools were thus opened for the study of new materials and their functional and aesthetic applications, productive methods and technologies, re-qualifying industrial production by preparing new Designers.

In addition to schools, applied arts museums were also opened, and the circulation of industrial repertoires was promoted through occasions for world-wide comparison, such as the World Fairs. It is in this context that experiences like Arts and Crafts were born, the movements of William Morris, and a school like the Royal College of Art; when it received the Royal Charter in 1967, its goals were 'to advance learning, knowledge and professional competence particularly in the field of fine arts, in the principles and practice of art and Design in their relation to industrial and commercial processes and social developments and other subjects relating thereto through teaching, research and collaboration with industry and commerce'. This definition still applies to the training provided by the Anglo-Saxon School that dialectically interprets the relationship between the concreteness of pragmatism and the creative dimension, between technical control of the Design and the tension towards experimentation, between controlling the functions of the artefact and typological innovation, and in formal research the aesthetics of the approach to formal research of *High-Tech* are expressed. With the *Werkbund*, then *Bauhaus* schools, in Weimar and Dessau, the experience expressed in the Modern Movement (MM) was born, permeating all of Europe with its principles and theories, before spreading beyond its borders, to the point where historians have called the methods and expressive forms used by this movement *'International Style'*.

matter and materials > In the beginning, matter was moulded to become a material, the fundamental element of all the manufactured items that have always surrounded us. Stone, clay and wood were used as instruments to erect and build objects, not just because they were accessible natural resources, but above all for their construction qualities, for their static strength, durability, workability, combinability with other materials, malleability, and beauty.

The focus was thus exclusively on natural materials, because they were functional, aesthetically pleasing, easy to use and readily available.

However, over time, the way we consider materials has changed considerably.

On the one hand, materials have interacted with craftsmanship, then with manufacturing techniques, modelling and modifying them to continuously adapt them to their own prerogatives. On the other, techniques — and subsequently language — have moulded materials, transforming them, expanding them, multiplying them making it increasingly difficult to distinguish between natural and artificial materials, or between the various categories and families. Indeed, in the 21st century a new way of understanding materials is being defined: they are no longer simply the constituent elements of a manufactured object; rather they are by themselves the form and function of the design. This has profoundly modified the physical and linguistic references that have always guided us in our day-to-day relationship with things, allowing us to interact with 'the material' of artefacts using all of our senses. This is how the change in our relationship with materials — which from passive elements are becoming increasingly active — has created an unparalleled emotional tension, expanding the possibilities of design; while destroying codified balances and points of reference, both those linked to what the material is and to what it can do.

In this framework, Nature has always been a vast and inexhaustible source of inspiration for possible future design scenarios. In the past, however, the main goal was to imitate Nature in order to ensure that a new material or product would be accepted. When plastic (bakelite) was first introduced, its creators looked to Nature in order to give all the products created using this new material — which was extremely practical and low-cost — a faux-natural skin. Decorations and fake grains or imitation skin, as well as colours and morphology of previous, codified 'Natural' materials were copied as closely as possible. The material's true nature thus took second place and instead was given a natural and familiar character; often, the material's very identity was sacrificed in exchange for a codifiable image based on individual memory. The focus was, in fact, less on the material's true nature, and more on the unchanging image it was supposed to transmit.

training and informing > In recent years, the commitment of the university system in the formation of the Italian designer with proposals up to the task, was passionate involvement in national experiences. Various methods and theories of design, but common goals: young designers in a position to know what is necessary to understand what should be known to make innovations to be successful in a global market.

In private institutions, which are important educational assets, only in very rare cases does a professional relationship between business and young designers exist, since the relationship between business and schools is based on mutual benefits of the business-school relationship, and since it requires the development of research into new types of product, proposals that the school takes up as its theme of the courses, so that at the end of the experience, the company may have a low-cost "inventory of ideas", without undertaking either the prototyping or engineering of proposals.

In the universities, on the other hand, the relationship with companies is aimed at conducting activities involving the training placements of the young in a direct relationship with the company, which the school supports with targeted tutoring to guarantee the student, from time to time, an entry into the world of work while protecting his/her creative and professional ability.

The results of the training are reflected in the ability of the young designers.

And the products designed by young designers for companies, are certainly a partial representation,

but significant nonetheless for the answers that we were seeking. And a selection of the output of young designers is inevitably partial and tendentious.
Just think of "The New–Italian–Design" the exhibition organised at the Milan Triennale (January 20/25 April 2007) curated by Andrea Branzi, and conceived and coordinated by Silvana Annicchiarico.
The event, in fact, has generated much controversy because of its cultural framework and choices.
Andrea Branzi, on that occasion, selected projects and products by young designers that confirm a thesis of his: our country is home to a constellation of creative workshops that express a set of new ideas that are valuable in themselves, insofar as the backdrop they take place against and spread its innovation.

difference and identity >   In the culture of design, difference and identity both play a strategic and complementary role.
In any case, differences and identity (identification/individuation) are influenced by the reference framework: the site, the historical and cultural context, materials and technologies, history and the culture of specific geographical and/or social contexts.
In the field of *design* or better still, in the field of *industrial design*, identity and difference of the object or product system, play different roles and, compared to the past, they fulfil a specific task which is often fundamental if they want to be successful in a global market.
It's worth mentioning certain aspects of identity and difference in the field of industrial design, for instance "gender difference", in other words, what tends to classify objects (that belong to the same grouping) according to the gender to which they are intended (man, woman, etc.).
Or, for example, the "difference" of products which, even if on sale all over the world, maintain their identity because they're linked to the specific geographical context of the market insofar as they are designed taking into account the differences that exist in material cultures: many accessories designed to cook and eat food or the furniture used in bedrooms or living rooms.
Or, again, the "difference" in the iconic value of the objects, in other words, how the success of some objects is dictated by their specific identity in a group of products with similar functions; their "identity" is more important than the quantity and quality of their technical characteristics (the iPod is an emblematic example).
However, it's more interesting to analyse how identity and difference become structured.
Designing industrial products involves creating objects for an ubiquitous market — with different histories and material cultures — and for groups with different needs. This is why industrial, large-scale production has, from time to time, tried to define a standard, in other words, a predefined characteristic or set of characteristics. By establishing standards, it's possible to sublimate all this "diversity" into a single "identity" in order to satisfy a global market. Or to "normalise" them and make them compatible with the needs of such "diversity".
In short, not products with many identities but many diversities, with a standardised identity

compared to so many diversities.

The last century was characterised by Henry Ford's standardised mass production and the so-called assembly chain. This production method manufactured identical objects and managed to cheaply produce many objects.

The assembly chain eliminated any sort of style — as an element of difference — in the objects. Identity and normalisation were necessarily integrated into the objects to ensure maximum production yield and penetration of the market.

We have lived through an age in which production has changed very radically.

Today, with the advent of post-Fordism and the new technologies used in manufacturing, a new model has emerged — not always the same, but versatile and capable of adapting to new needs.

Today consumers can choose from a variety of different products.

Naturally, this increases supply and reduces production costs.

In some cases, the company can wait for individual clients to make a request and then satisfy him by assembling the right modules. By doing so, he can economically produce batches of just one unit. Not all consumers realise they are making a sacrifice, but if the extent of the sacrifice is limited, then the consumer is more than happy and returns the favour by being loyal as well as being willing to pay a premium price.

Currently, many big companies are moving in this direction: for example, the multinational *Nike* wants to ensure that every client is able to personalise its shoes at a price that is almost identical to a standard pair. The Smart car is marketed with a whole range of parts that permit diversification, over and above the normal bodywork options (colour, interiors, type of wheel rims, etc.), which is, in fact, considered as a combination of interchangeable parts. The client can assemble the various parts of the frame to create his "own" special car.

the event of the exhibit > The international circuit of design fairs and exhibitions is an excuse to examine the relationship between the added value of a product, its innovative content, companies, designers and the influence of trends and the market on the characteristics of products.

These fairs/exhibitions are created to endorse companies and products; window-displays that on a regular basis propose new styles for industrial products, new types of objects, or are increasingly full of functional and technical content.

Initially created for "commercial" reasons, they also focus on experimentation and the challenge of innovation. These events dictate the success of designers and companies as well as finding markets for certain products.

They are boosted by the strategic contribution of communications media, specialised publishers and exhibit design.

Specific interest groups and businesses have developed thanks to these events which, amplified by the media, have reached beyond the companies, designers and new products and intercepted the enormous resources used in communications and the spectacularization of the event to attract the public at large and make them participate in the increasingly intense calendar of engagements that have become permanent events.

These events have adopted the methods used by fashion shows all around the world, events in which communications and presentation during shows has a predominant pre-textual function. Even in design exhibitions, media communications and the design of extremely creative and imaginative exhibits are even more important than the efforts of exhibiting companies to renew their productive processes, product types and performance characteristics.

The true performer is the exhibit and not the presentation of the product that isn't always innovative.

The design of the exhibit can be an excuse to represent a theme and/or an emotion.

In Earthly Paradise, an event of the Fuori Salone (Milan) in April 2003, Mendini designed a real "master plan" for a system of sets and backgrounds in which each of the famous designers designed freely, but tackled the specific theme and Mendini's interpretation expressed in the relational spaces and overall structure as well as in the overall communications project.

The places where we look after our body and our feelings become an earthly paradise, a place of well-being for all our senses.

The exhibit design can be a chance to represent the immaterial.

For instance, at the elegant Lexus exhibit at the Triennale, Kazuyo Sejima and Junya Ishigami tackle the immaterial and the absence of spatial reference points. In fact, the car is displayed by eliminating its presence in a milky space in which every material object disappears in the mists and absence of colour.

The design exhibit may be the chance to "monumentalise" cult design products.

The exhibit design can be a chance to experiment with stage effects.

For instance, in the 2005 Fuori Salone (Milan) where Vitra displayed its products in an exhibit designed to encourage the visitor to focus and concentrate on the work of Ronan and Erwan Bouroulecq. In fact it is the exhibition area that is highlighted, while the collections of products seem slightly out-of-focus and of secondary importance.

The exhibit design can be a chance to enhance a surprise effect.

For instance, the big Milanese Bruco di Recapito, a pneumatic, flexible, movable structure, gigantic in shape and size: a totem of immense communicative power on top of a building, with its face turned down towards the public. It is a playful and inspiring structure, very visible during the day and lit up at night, that fascinates people and invites them to enter. It acts as a meeting area for thematic exhibitions and displays for companies, designers and institutions.

The exhibit design can, finally, be a metaphoric Representation, for instance 100% Design in Tokyo where for over a year the exhibition areas were set up in containers. The assembly or organisation of a container acts as an urban passage, a metaphor of the warehouse and global trading of goods. The way the containers are arranged tends to build a fragment of the city of the exhibit in which the containers display a catalogue of exhibit projects that highlight the role of designers, rather than the products on display and illustrates all the possible approaches towards the culture of exhibit design.

virtual architectural complexes > The introduction and transmission of information technologies in the fields of design and communications has radically changed the work of graphic artists.
In the past, communication designers used to graphically and typographically manipulate signs and letters to express various kinds of communications using two dimensional styles.
Today, they use a new, three-dimensional, dynamic and interactive environment to express themselves: the graphics are no longer based on a single structured representation, like a poster or the page in a book or magazine.
Communication designers are asked to work with tools, for example the web, that interface with users and involve a series of levels that transform a traditional sheet of paper into a place that has multiple access points that dynamically lead us down many paths and into the complex communications and information systems typical of website navigation.
Apart from the text, the kinetic composition of images and the use of video sounds modify the traditional methodological concept used by traditional graphic communications.
At present, the most important expressive fields used by communication designers are the complex virtual architectures of websites: environments with several levels, three-dimensional representations and three-dimensional scenarios in which every element of style is affected by the circular processes typical of the web, by the use of techniques, software and tools that allow interaction with the final users of the communication, by a system of superimpressions of transparent layers characteristic of each hierarchical level of the messages and communicative communications.
The communication designer has to come to grips with the fact that distances don't exist on the web, that the images are launched into a universe of different cultures, turning a globalised style and level of communication into a shared piece of property.
Today, surfing the web is the new way of interpreting communications. The designer has to tackle this problem and allow the person who reads his messages to decode the "information to be used" from the "informative contents" of the site.

it is design > The environment around us, made of artificial products — material and immaterial — is the result of the planning and industrial production of objects of countless shapes, technologies and dimensions, and addressed to various uses.
The design, in its varied and multifaceted forms, determines habits and new behaviours within interpersonal relationships, affecting the material outcome of contemporary society: through the innovation of products, their production processes and planning.
Within this context, the concept of design should be considered, not so much as in semantic terms, as within the reality of project-making itself; this way, it acquires a new meaning as a conscious planning activity, as a virtuous feedback, able to formalize objects and/or services, new products for old and new needs.

Therefore, we do not only refer to the kind of design that's recognizable within the context of a more aware and rather elitist part of contemporary society, but to a design that refers to normal and daily aspects of living: a design intended as a planning activity applied to convenience goods or to developing enterprise sectors, to the complex scenario of planning requirements within advanced industrial societies.
This is the *design* of new products, conceived in such a way that, at the end of their life-cycle, their existence/use can be extended under other forms.
It's the design of new products that are programmed as systems of interchangeable elements whose application and assembly permit the realization of numerous and different models.
It's the *design* of new products to whose basic functions additional performances can be added, thus increasing the global complexity of the product itself.
It's the *design* that marketing strategies use, as Philip Kotler writes in Marketing management

"....marketing is the business function that identifies current unfileed needs and wants, defines and measures their magnitude, determines which target markets the orhanization can best serve, and decides on appropiate products, services, and programs to serve these markets. Thus marketing serves as the link between a society's needs and its pattern of industrial response ..."

It's the *design* that becomes a resource that's able to turn social observation data into a synthetic representation and that, through creativity, suggests something not yet existing to satisfy new needs. The *designer* becomes a complex figure whose skills affect a scenario that's characterized by professions as well as production and commercial organisations strongly depending on machines, information and automation, technologies in continuous evolution that require constant updating.
The new concept of *design*, traditionally only recognized in its role of shaping products, refers now to the complexity of industrial planning as a whole.
Education and training become therefore a permanent condition, which requires a learning method able to transfer advanced know-how but also the ability to understand the complex relations between demand, offer and ability to respond; the knowledge and understanding of skills and methods to manage complexity; the competences to analyse the demand and to know now to translate it into a suggestion of innovative objects.

the language of the object > In essence, the industrial product's belonging to the ethics and the culture of globalisation begins right from its beginning.
The project sublimates the concept proposed, fixed in the drawings, simulations, models, forms and uses of the object; it contains the system of relations between it and its end-user, determines the points where the object is sensitive to the touch, sight and hearing of those who will possess it and will be able even to smell and taste it; it combines a complex system of components, mechanical parts and electronic circuits into a simple product, establishing even its production means and

timeframe.

There is no doubt that it is the project that defines the object's language, and style, and thus its added value: contemporary languages that seem fragments of a system in continual search for a new identity; notions and forms like splinters unable to embed themselves in established structures, to transgress in order to assert that desire for hegemony typical of the avant-garde; styles that express that condition of cultural acceptance based only on the continuing search for combinations of exotic and appealing references inherent to what we could call a *Global Style,* a diffuse language that is shaping the scenario of our daily lives, and taking the entire planet as its spatial reference.

The production system is that which has more obviously revolutionised the operative procedures and logic itself of the methods of traditional work. It is certainly the phase in which technological innovation has had the most impact and, therefore, it is the phase upon which ethical, social and economic implications weigh in a more significant as well as complex and contradictory way.

An object is die-cast in one place and its assembly or warehousing do not even take place on the same site. It has already been estimated for ten years now that of the price of an automobile sold by General Motors 40% is spent in Taiwan of China, Singapore and Japan for small parts, 30% ends up in South Korea for assembly and finishing, 17.50% in Japan for high tech components, 0.75% in Germany for the design of mechanical parts, 0.25% in England for advertising and marketing and 0.5% in Ireland and Barbados for computer processing.

Today, with the introduction and diffusion of remote controlled computerised production plants and robotics, which practically exclude margins of error, perfection to the millimetre has been achieved even in the production of much less complex objects, and production processes are more and more separated; on the other hand, margins for product improvement are enormous with the possibility of assembling components selected from the specialised production of homogeneous sectors.

If, on the one hand, this spreads work and wealth, on the other it slows the integration between excellent decision-making skills and the technical skills operating peripherally, and accentuates inequalities. Economist and philosopher Amartya Sen, in his book *Globalization and Freedom,* observes that inequalities are the principal source of doubt about the planet's economic order. He points out that, even though our world is richer now than it has ever been before, it is one full of tremendous privation and disturbing inequalities.

prospective scenarios > We are seeing daily, with our eyes and with prostheses such as eye-glasses and contact lenses, thousands of produced images: those in the newspapers, from the windows of our homes, from our moving cars, on the luminous liquid crystal or plasma screens of our monitors.

Our olfactory sense is stimulated silicon and glues, by the new odour of plastics, by the dry air of airconditioners, and a thousand particles modify the scents of the seasons recalling the aromas of distant continents.

These objects and contexts meet in an intermingling of form, image and language that concerns the processes of their production, the relationship between the objects and defined contexts such as the city, the metropolis, and the man-made and natural territory, and, finally, our bodies, minds and sense apparatus. In this scenario the design and production of objects, and the control of their life-cycle up to the point of their obsolescence, represent moments of synthesis and complex decision-making that see the investment of a multitude of technical competences, new forms of artistic expression, new trades and new social and ethical content.

Analysis of the industrial product, therefore, allows for a greater comprehension of our times by reflecting on the relationship that the object establishes with a market free of territorial or traditional barriers, and with the confirmation of the material practices of the so-called global village.

It also allows us to look more deeply at how the industrial designer's task has been complicated in its concern, no longer solely with the product's aesthetic appeal, but also with more articulated and specialised issues as well: product design, component design, fashion design, lighting design, management design, strategic design (the design of innovations in process and product) ecodesign (the control of the product's eco-efficiency), the design and planning of cultural events, museum design, the design of products and product systems for special areas such as historic city centres or archaeological areas; the design of multimedia and interactive visual communications, with appropriate languages, tools and the necessary and possible technologies that go into graphics for publishing, packaging and for co-ordinated, animated and synthetic images and the interfacing of computerised networks.

The designer works in sectors where technology is utilised at its highest levels, in sectors whose production strategies look to the future, creating technical objects, objects for use and objects that do not exist at all, anticipating new and unimaginable performance possibilities, new materials and unimaginable applications of them, for a market that unites the demands of the men and women of the advanced industrial society with the demands of the men and women of developing societies.

[2]
p.45

Conceiving consumer goods

new technologies
from Il design è ubiquo | Lectures 1, 2013
colour
from Il colore racconta tutto il design, DIID 53|2012
teaching methodology
from Design fra identità e diversità, DIID 42–43|2009
environmentally sustainable technologies
from Ecodesign e nuovi scenari, DIID 41|2009
connections
from Dipendenze tecnologiche, DIID 39|2009
ambiguity of meaning
from Materiali e natura, DIID 38|2008
training experience
from Designer After School, DIID 32|2007
market demand
from Design off Limits, DIID 26|2006
spectacularization
from La spettacolarizzazione del design, DIID 18|2006
communication
from Frontiere della grafica e della comunicazione visiva e multimediale, DIID 16|2005
deception
from Contaminazione, trasferimento, omologazione, DIID 14|2005
thinking materials
from Sincerità e ambiguità dei materiali, DIID 13|2005
magnetic cards
from Design is everywhere, DIID 7|2003
useful obsolescence
from Trasferimenti di senso, DIID 5|2003
playful appearances
from Tool Toy, DIID 1|2002

It often happens that elementary things escape us. This is something we all know: concentrated on a phenomenon as a whole, we often distance ourselves from its details. In this sense, the risk the design system runs lies in the fact that in showing oneself to be free and inclusive as is required of the knowledge of diffuse design, it underestimates its natural mandate in the name of too many assessments — ante-, post-, meta- and trans — that precede the design choices. Its final task, after all the premises and articulations that accompany its phenomenology, is to design artefacts and to do so in connection with its social mandate: interpreting man's need to represent his material culture.

new technologies > The axiom "Less is More" then takes on new meanings in comparison with its origin: it becomes the search for new equilibria between artifice and nature, through a lesser utilization of matter and energy, and more attention to the conception of eco-friendly artefacts; it becomes the ability to control the most miniscule tolerances in the working processes, and stimulus to subtract figurative elements that are redundant and invasive on the formal level; it becomes the impulse to design and make sophisticated technical objects, small and intangible, rendered intelligent in their functions by the technologies of microprocessors, and quick in their interaction with people, who can use them with simple, ergonomic gestures, and without requiring in-depth knowledge; lastly, it becomes a research opportunity for physically reducing things for future scenarios, through the application of nanotechnologies, which is to say a form of extreme shrinkage. Technological innovation impacts the evolutionary dynamics of technical objects: until today, these objects have been, and still are, material products that, like extensions of the body, carry out — in a refined and effective way — specific functions already included in man's sensory apparatus, or expressed by his relationships with natural needs, or by his relationships with nature, or by his system of social relations, or by his behaviour in all the activities he is involved in.

New technologies act upon biological processes, forming genetically modified organisms, which is to say making it possible to extend the concept of artificial and industrial product to food products: tomatoes, courgettes, and other foods, once the result of a natural development cycle, take on colours and sizes designed to respond to a market demand in the most satisfying way possible. New technologies make artificial products smart, so they can interact and dialogue with man, interpret his moods, listen to him from afar, urge his well-being, or form bonds between one person and another. New technologies make it possible not only to extend artificial organs outside our body, but to directly integrate new functions into our own bodies, like seeing in the dark, speaking or hearing underwater or from afar, or seeing beyond real dimensions. A condition is once again advanced: of man and his relationship with science, and the unexpected technological applications that are certainly not entirely unknown; however, never more than today, the potentials of technological innovation encourage us to overcome every imagined limit.

colour > The use of colour in industrial products. The breadth of the topic, the many variations on the theme, and the wealth of literature about the associated phenomena are truly staggering. Paradoxically, such vast quantities of information do not always work in the favour of researchers and scholars. In order to overcome this stumbling block, the decision was made to put together some reflections on the unconventional aspects of colour in the design of tangible and intangible products, in both the present and the past. I hold three of them to be particularly noteworthy and of great personal interest. The first aspect confirms that colour is an essential part of the experience of the designer. As Henri Matisse said, rather than drawing, it is colour that offers liberation. In other words, all of the possible chromatic changes unleash a broad spectrum of sensations. Therefore, when we reproduce coloured signs and decoration, we are doing more than embellishing things with a range of hues as one would with a colourful seasonal garment. We are working with dynamic properties that continually drive relationships between places, objects and people. These ties can be measured in orders of synaesthetic magnitude and they affect sensory interaction, establishing connections between the invisible and the visible. This same junction point is also the source of the complex selection of metaphorical and linguistic devices that express colour.

The mechanisms in question have been generated over time in accordance with different cultural grammar systems. Sottsass maintained that colours were words. Given his history, we might imagine that they served to write visual documents that were as ancient as they were modern, meaning that they were capable of dialogue with everyone. "Visual writing" is also a key topic in the experience of Alessandro Mendini. As it is on the surface, it makes the surfaces themselves descriptive, communicative, and therefore sensitive. The second factor seems obvious, but nonetheless it is necessary to dwell upon it: colour is a design and one designs with colour. It is a tool and a means of producing the complex rhetoric of visual "things", which includes tones, symbols and abstractions which can be seen in the surfaces and volumes of the products around us. Back in the time of the Bauhaus, colour had already been promoted to the status of a new design category thanks to sophisticated contrasts between primary colours which — together with the geometry of forms — served to balance functions. However, one of colour's most comprehensive displays has been given in Italian design. Whether it is a question of hyperbole or chromatic prohibitionism, natural or artificial colour, history has shown us that for those responsible for the relevant choices, the colour of objects is a fundamental issue and never a secondary one. All of this is even more apparent in the modern world, as ranges of colours can be developed thanks to pigment chemistry and the reproduction capabilities of digital techniques.

teaching methodology > Bauhaus attributes to Design the task of directing industrial production towards diffusing objects of formal and functional quality that can contribute to the aesthetic and moral education of the general public. A didactic method based on experimentation was adopted to train Designers from a technical and artistic viewpoint, teaching them to work in teams comprising the many skills needed to provide technical solutions to complex functional and technological problems for the production of industrial artefacts, but also to establish transversal relations between Visual Arts, Design and Architecture to formally define products.

It's no accident that many masters of modern architecture have received training in Design; it's no accident that there have been many stylistic analogies between Architecture and Design; and it's no accident that in Europe many Design Schools are still integrated with Schools of Architecture or with Visual arts schools, or both.

And Design Schools which belong to Polytechnics or Engineering Schools are rooted in that same experience, even though they have focused primarily on organizing industrial production in terms of technical innovation and the services provided by artefacts.

The great masters of this experience, or those related to it, are Marcel Breuer, Eilleen Gray, René

Hernst, Mies Van Der Rohe, Jean Prouvé, Le Corbusier and Walter Gropius, who also built the identifying traits of the French School and of that of Middle Europe, and who also influenced the identity of Design expressed by other regions outside European borders where *Bauhaus* teachers and former students worked with extraordinary determination, never losing sight of that social commitment that was the ideological foundation of *Bauhaus* and of the MM.
Starting in the 1950s, the Ulm School helped to enrich the content of Design, including theories of communication and information in Design studies, and the management and control of the immediate use of the object in its operative field, linking it to its processes of obsolescence.

environmentally sustainable technologies > However, we are already aware of the content of the numerous experiments in which designers and companies attempt to propose sustainable products whose forms and functions optimize use and operating times, while the materials utilized and production processes allow moderate consumption of natural resources.
The field of experimentation is undeniably full of ideas that are definitely more evolved than those a few decades ago in that there is a greater focus on their appearance, but despite this they are unable to move away from formal and typological residential models bound to a concept of urban and domestic space that is now antiquated. Indeed, it seems that designers still consider technologies for sustainable living spaces to be invasive devices in buildings, as in spite of everything they stick to traditional views of space and its formal structures.
The field of eco-sustainability simply has to face up to new technologies, both the tangible ones and the intangible digital ones that I believe are bringing about a genuine cultural revolution. It must therefore push for a new approach to design and create a fresh concept of living. To give an example, even the architect Mario Cucinella — who is perhaps one of the most interesting sustainability players on the European scene and has gone so far as to conceive a house which is not a "machine" that consumes energy but rather one that produces it — does not seem to give his bold idea an architectural shape that reworks its typological and aesthetic paradigms.
This approach is widespread and concerns more or less all of the design world, including that of industrial products. There are intelligent monitors whose brightness is reduced when sensors detect that the user has moved away, extractor fans that direct dirt to a device where it is compressed to reduce the space it occupies, mobile telephones that are made of recycled materials and have energy saving functions, refrigerators with special panels to make their insulation as much as 20 times more efficient and the emissions of cars will soon be reduced to zero. There is a wealth of extraordinary technological innovation that is nonetheless impossible to perceive in the forms or interaction with the user. No concrete shape has been given to the new vision of the world that the system of sustainable products should really be supporting. It seems that even though we are dealing with factors of technological innovation arising due to the need to save energy or design alternative forms with digital support, we are not fully aware that we are moving towards a new conception of the world and a new relationship with space and time. Yet the designers who are handling the new

technologies find themselves in the same situation as their predecessors at the end of the industrial revolution, such as those involved in Bauhaus, who tried to give some meaning to the shapes of the technology of the time, as part of a new aesthetic revolution.

connections > Today, it is possible to design environments in which man can be immersed in the future "Touch Radio" (Radio Tattile, whose prototype was created by a team of designers of the Interaction Design Institute of Ivrea). *Tune* is a "Touch Radio" which you enter to explore emotions, visual experiences, space morphing, accompanied by the listening to sounds and music and where lights features show images changing upon the choice of the different channels. At the Ulster University a prototype of IMP (Individual Memory Projector), a kind of memory box of our memories, was developed. It is shaped as a sphere that opens by rotating it and turns into a small compact projector interacting with man bringing him back to his memories, which are projected on any wall as family movies and photos in association with sounds and smells of the environment recalled.

Researchers at the Carnegie Mellow University of Pittsburgh created a prototype of *The Hug*, the "love cushion" that makes the person you phone feel sensorially closer. The filled cushion is actually a robot with rounded shapes suggesting a head and two feet, which is equipped with software to recognize voice and other signals. During phone conversation, it features your answers and speaking into the microphone, and reproduces human interaction through lights and warmth (radiated thanks to special thermal fibres) and through the vibration pattern generated in association with the sensory data sent by the other Hug.

Through electro luminescence, OLED (Organic Light-Emitting Diode) technology allows users to have flexible, ultra-thin, ultra-light and fullcolour displays provided with an extraordinary level of brightness and contrast, which can be used as PCs, TVs, mobile phones to be rolled up after use. Interaction Design is a potential field of application that imbues with technologies the products and the systems of artificial products surrounding us. Though they are hidden, technologies are now a constant presence in our daily life and influence our behaviour, propose new dreams, stimulate our senses and our entire perceptive system towards new forms, open us to new experiences without any cultural mediation, spur us to interact with the artificial world to change it and to change our emotions or to fulfil our needs.

Interaction Design, as it plans new ways of interaction between man and artefacts to facilitate their use in relation to software technologies, thus requiring a complex system of different expertises : software and hardware technologies and engineering, the languages of the particular forms of communication, design culture and culture of the industrial manufacturing of artefacts, sociology, ergonomics, psychology.

The aim is to make the use of the artefact as accessible as possible and to turn it into a social and cultural enrichment factor for man.

The products of Interaction Design cannot be separated from technology and they can only face both the immaterial and virtuality as well as universal iconic languages. With man being the target of its products.

The use of increasingly developing technologies necessarily bound to the products of Interaction Design makes us think over the implications of such a strict dependency.

What is said above is even truer when considering that the products we are talking about are mainly immaterial: they come to life and gain a reason of meaning thanks to the action of man who interacts with them. It is man's action, and in some cases his only presence, to reveal their functions and uses and it is again man's action that may even determine their form, possible figurations, sound, scent, etc. However, man only apparently plays an active role as the object contains software programming and managing any of the possible configurations of the artefact.

ambiguity of meaning > Over time, we have begun using new materials to express their truer identity that they have won back. It is no longer true that wood expresses refinement and plastic vulgarity. Rather, the language of wood reflects the straightforward representation of the work of humans and of craftsmen with their tools that cut, plane, smooth and polish; and the language of plastic the hermetic and ambiguous representation of the transformation of chemicals and of its magic in which humans in their ingenuity invent processes that generate materials and allow these processes to express themselves even after the object has been produced, when the material is modified through its interactions with humans, time and the environment. Over the years, this type of "ambiguity of the surface" has been opposed to a sort of "ambiguity of meaning". Here, designers also use new materials to amaze us and to express non-explicit — rather, vague and changing — languages, in order to fool the senses and play on our comprehension of their physical consistency. They try to create new relationships between what is artificial and what is natural, between useful objects, technical objects, intelligent objects, people and the environment. And they sometimes oppose the straightforwardness of the languages of traditional materials with the misleading vocations of new materials which are constantly being made lighter, changeable, transparent, technical, luminous, reflective, cold, chameleon-like, and deceiving. Today, however, when we look to Nature, we seek to imitate not only its aesthetic aspects, but its more intrinsic characteristics, successfully designing the sensory imprint of the material and of the product, and also seeking to make it bio-inspired, bio-mimetic and even alive. So, we focus more on the material's natural character and on its intimate essence, rather than on its superficial appearance.

training experience > What is demonstrated is a capacity to measure itself against the complexity and richness of the system described above, but it seems that it remains a bit on the sidelines to quite measure up to those areas where tradition prevails and where products presuppose contents performance and advanced techniques. Or to push for experimentation in "style" with an attention to the manner of language without ever hardly being able to open up new horizons.

It is a scenario which calls upon the schools to measure themselves more effectively against the new frontiers of science and technique to translate them into new technologies for new products that improve the quality of life of the people they are intended for.

It is a scenario where young designers in only a few points of excellence express the ability to look at the complex network of factors in innovation: those who have the ability to analyse the phenomena of and around design beyond that too glossy and worn-out concept, to extend the concept of design beyond the traditional notion.

For young people, the school must provide more tools for a training experience that allows them complex overviews, as is required in the design of industrial artefacts which express the highest levels of technological innovation, which set strategies for production for the future, which develop technical objects, foreshadow new and unimaginable uses, new materials, the application of these in a market that unifies the demand men and women in the advanced industrial societies to the demand of men and women in the developing world.

market demand > New technologies affect biological processes creating genetically modified organisms. In other words, we can use them to extend the concept of an artificial and industrial product to foodstuffs: tomatoes, courgettes and other foods, once the result of a natural cycle of growth, have colours and sizes that are designed to satisfy market demand.

New technologies make artificial products "intelligent" so that they can interact and dialogue with us, understand our moods, listen to us from a distance, make us feel better or mediate between two people.

New technologies not only allow us to maintain artificial organs outside our body, but to directly integrate new functions into our body, such as seeing in the dark, speaking and listening underwater or at a distance and seeing beyond reality.

spectacularization > Designers are no longer asked to design only functional products, but above all to design attractive objects that the consumer can't resist buying, objects that have contents and meanings that are symbolic rather than real or that emphasise a dematerialisation of the products that become communication, a complex style with its own rules, its own grammar and a syntax that doesn't only focus on satisfying needs. If, on the one hand, ideas appear to have no material bonds, becoming once again free thought and pure signs, on the other, the social role of the designer appears to inevitably dissolve into a world of lights and sequins. It's short step from the spectacularisation of the product — or even of just the project and the idea — to the spectacularisation of the designer and — in a completely consequential cause and effect relationship — the latter seems to adjust to this logic.

Real icons of the star system, famous protagonists, but also skilful managers of their own media power, designers are always clever strategists when they have to communicate the product and their own image. The designer is "media," he is the attraction to touch, see and listen to: stars among the stars and masters in this field, Philippe Starck — whose face is also a brand.

communication > They also train students to be a designer capable of using all available communication tools that range from paper to monitors, from the analogical to the digital, exploiting a tradition which, on the basis of one of the oldest forms of visible representation of images and words, saw the beginning of printing in the mid 15th century and the advent of internet in the 21st century.

We should take a moment to talk about the rather tricky issue of the relationship that visual and multimedia communications and its specific field of elaboration, experimentation and scientific research has established in recent years with industrial design and its own specific field of elaboration, experimentation and scientific research, in a common disciplinary framework of reference on the question of design.

In other words, according to the declarations of those working in its own field of reference, the scientific and disciplinary contents of this "area": involve the theories and methods, techniques and tools of the design of industrial products — material or virtual — that apply to its productive, technological, constructive, functional and formal traits and to the relationship that it establishes with the spatial and environmental context and with the world of industry and the market. The nature of this product (ranging from assets and capital goods to long-lasting consumer goods, from communicative, relational, interactive artefacts to relational and service structures) and its complexity (from materials and semi finished products to intermediate goods, components, final products and integrated systems of products, communications and services) involve just as many design methods and techniques as an interdisciplinary procedure which, interacting with the various goods sectors and fields of production, give rise to specific research fields in continuous flux ..." In essence, they involve knowledge and skills in related fields, transversality, knowledge that is specific to theoretical speculation, complete with the techniques and tools used in design projects and experimentation.

So if it's true that visual and multimedia communication theories and practices are useful and should be carried out or gradually be integrated into the theories and practice of the design of everyday objects as well as being tested against a design culture which is both the 'joint' and compulsory common reference, it's also true that the specifics of each field is indisputable.

That's why I believe that if it's appropriate, even necessary, for these two kinds of training to be part of the field of industrial design, it is also necessary to emphasise that experimentation and scientific research of both fields should be kept within specific boundaries: limited boundaries, with common borders, but with their own specific disciplinary statutes.

deception > Even if the product is industrial, it is designed for short-term serial production, market segments in which the formal contents of the object and its aesthetics are in a state of constant flux, while the functional parts remain the same.

This has produced a relationship of transferral with architecture: in fact, architecture now rapidly benefits from the contents of a research developed to produce industrial artefacts.

By using the technologies experimented in a technical object, architecture is increasingly becoming a "machine" as complex as an industrial product.
In fact, while buildings are gradually dematerialising they are also full of all the electrical and mechanical parts that improve their performance.
They have communicative and symbolic values which, like an industrial product, become added values: however, the latter actually shorten the lifecycle of the architecture and exponentially accelerate the obsolescence of the building itself.
The widespread use and development of the much improved, almost perfect frame structure has led to the separation between the casing and the load-bearing structure, creating a sort of skin around the skeleton of the construction.
Apart from the injection of technologies, building methods, performance and symbolism into these "metropolitan monsters," all the traits of an industrial products, simulacra of technical artefacts, have been shifted onto a larger scale. Even the relationship with the site — that has always been one of the cornerstones of architecture — is much less important.
Neither the shape nor the materials used have an organic relationship with an object's ultimate function. Ambiguity and deception are endemic in the design of objects and architectures: their ultimate use is secreted and advertising highlights this ambiguity.
Objects and buildings become more and more complicated because of their communicative nature: buildings and the city assume ambiguous, indefinite shapes turning into ephemeral places of communication and advertisement. The more highly innovative the technologies, the more artefacts and architectures become ephemeral.

thinking materials > Our relationship with the materials in the things that surround us provokes an emotional bond with our historical memory: something a designer sublimes when designing a product. After all, materials are the "literal components" of all artefacts, because materials make them real. They are the ingredients that man has used or uses to make products.
Over time, materials have been worked by first by artisans and then by machines to shape and change them, continually adapting them to the producer's needs, but materials have also shaped and changed artisans and production techniques.
Materials are used in art and are the premise of the arts; their power lies in being able to conjure up a whole culture with their immediacy. Materials such as stone, clay and wood that were chosen as tools to build objects. They were used not just because they were available, but because of their characteristics, their static strength, durability, workability and also because they could be combined with other materials, bent and turned into beautiful objects.
Jean Nouvel's design of the Tour de la Défense in Paris is the emblematic theorem that intelligently represents — with an amazing flair for synthesis — the double language of the ambiguity and honesty of the materials used. The perfectly cylindrical Tower rises up in all its "sincerity" out of the bowels of the earth with the natural irregularity of a block of rough, black granite. It soars skywards,

and beyond the tenth floor the granite appears to be smooth; its shiny surface reflects the light and colours of the sky in a hazardous "vagueness." When further up it dissolves into black glass, the deception is complete: granite and glass have become a mixture that makes them dematerialise. Even further up, they appear even more inconsistent when the serigraphed, opaque glass makes everything more evanescent. Then finally, up among the clouds, the glass reappears in all its "sincere," transparent, material consistency, blending with the sky.
New materials provide remarkable opportunities in experimentation in all fields of art.

magnetic cards > It's the design of items which combine immaterial relationships through material products: like the smart card, an object that — minimal in its shape, maximum in its content-connects us to several links and transforms us in a computational algorithm.
An intelligent card made of plastic, on which the Japanese Kunitaka Arimur inserted micro processors that were able to elaborate and hold data.
These cards, like real electronic security boxes, aren't only used as credit cards, but also as phone cards, as keys for security doors, as cards containing personal data, as access cards to digital TV programs.
The smart card represents one of the most important factors of transformation of our behaviours.
It's interesting to see how much value magnetic cards have gained in each of their uses — as phone cards, badges, credit cards — within contemporary people's "kit of tools": a rigid piece of paper, a "card" indeed, with specific dimensions more likely required by their interaction with machines than with people, that however inevitably influence their set of accessories (i.e. wallets, cases, pockets).
Also interesting is the interaction magnetic cards create between service providers and users.
Think, for example, of supermarkets and petrol-stations costumer cards to collect points at each purchase; or identification cards (i.e.in Prada stores) which collect information about each customer in order to customize their offer as much as possible. The smart card even becomes a tool to be used for business strategies, also related to each single product.
It's about the design of new products for which the accurate choice of materials and technologies is fundamental, because their characteristics have to respond to sensorial needs, like contact, pleasure and comfort, more than image itself. It's the case of Technogel, a biomedical material, used by Philippe Starck and Gaetano Pesce, to create specific objects whose main characteristics are tenderness, softness and protection. Such products satisfy the need for entertainment, pleasure, emotions, that is the sense of wellbeing defined by marketing experts like Faith Popcom as "cocooning".
We talk about a design that avails of methods that, with the use new informatics instruments, can now interact with production processes in tighter connection, to the point of inducing real transformations in planning logics and practice.

useful obsolescence > Another way to manage the life cycle of an artefact is to slow down its decline by changing and upgrading its technical components, in order to make the performance compatible with developing technology; cannibalizing, that is to select or take two or more obsolete products from still working parts in order to make an efficient one; replacing with a new and upgraded product and management of dismissal carried out by the providing company.

Nonetheless, a product may be reused for different performances according to tradition when its functions become obsolete. How many times have you seen food cans used as flower pots in gardens and an old shoe as a vase? How many times have you seen newspapers used to wrap food or make caps?

Materials and objects are often reused by people with no or little resources. For example, toys made by children in the streets with dismissed materials: skateboards built with old wood panels, coaches made of cans and wire, bicycle wheels turned into hula hoops and plastic bags that become kites.

An artefact whose functions have become obsolete can express itself in a new dimension, more as a superstructure then a structure. Its nonsense creates new scenarios, justifies the new aesthetic contents and makes it almost poetical.

For example, former drive-in cinema screens became huge pictures standing out against the landscape: on the dirty surface of the screen the old structure designs a Cartesian plane, and the sky pops up through the ripping of the screen. Apparently, they have always been there.

Abandoned gas stations, isolated steel warehouses, silos and tankers in the middle of the country appear like huge works of art built in the sky, evoking functions and traditions from the past, which sublimate them as archaeological sites. Like sculptures in an artificial world, turned into natural elements as a part of the landscape, we are inspired by old and abandoned rusted freight cars, whose old structure reminds us of the long trips and the faded away posters of the companies and products transported.

Yet, artefacts are not always abandoned when their functions become obsolete. The added value of the product is often acknowledged in its aesthetic content and representative value.

playful appearances > We have analysed the tendency to design and produce objects of daily use as if they were toys.

Objects which, with a sense of irony, conceal their true use and function behind a playful image-the object as ToolToy.

This is a stylistic trend arising from an acknowledgement of the communicative content of an object with added value.

The attribution of value to the object, in fact, is no longer an expression of the balanced relationship between its aesthetic content and its performance, but lies rather in its formal expression.

Industrial products of recent years have gradually dematerialised through the use of extremely innovative technologies and functions, and have become the accumulation also of the most advanced image research, achieving a level of performance that is neither useful nor practical, but really quite superfluous with respect to primary needs. There are so many objects that could serve as examples of this, but a particularly effective one is the watch.

With the application of the principles of the spring mechanism or the pendulum, the goal of absolute precision had been accomplished, and precision was precisely the parameter that guided one in choosing a watch — little did it matter if it had been designed by an unknown and was of scarce formal value.

Then, with the help of electronics and the diffusion of printed circuit boards — tiny components upon which the entire functional process of an electronic apparatus was industrially printed — a watch's precision became impeccable. Something else had to come in to give the object its added value: its image value, its design, its symbolic meaning.

Things, objects and products

complex paths
from Il design è ubiquo | Lectures 1, 2013

without interruption
from Naturalezza industriale, DIID 55 | 2012

urging ingenuity
from Il giocattolo come oggetto del desiderio, DIID 54 | 2012

architectural object
from Off scale and contemporary landscape, DIID 31 | 2007

strategic value
from Design for health, DIID 27 | 2007

at the service of sport
from Performance & Design, DIID 20 | 2006

anonymous but industrial
from Mass design o il potere dell'oggetto anonimo, DIID 15 | 2005

between objects and contexts
from Contaminazione, trasferimento, omologazione, DIID 14 | 2005

sincerity and ambiguity
from Sincerità e ambiguità dei materiali, DIID 13 | 2005

technological innovation
from High technology, DIID 9 | 2004

hybrid products
from Ethnic & technical, DIID 6 | 2003

design museums
from Trasferimenti di senso, DIID 5 | 2003

upgrading
from Global design, DIID 3–4 | 2003

With absolute simplicity, when we recount the artificial world, we represent it inhabited by things, objects, or products, indifferently: terms often used with equivalent meanings, as synonyms to describe what surrounds us. But they do not entirely overlap — they do not in the literary language, and they do not in the language of design. Despite this, they represent a scale of values in the design's dynamic and authorial dimension, and in the passage — also rhetorical — from the randomness of events that gives life to our material goods, to their voluntary determination. This is a deduction that establishes the nature of industrial design, which is to say transforming things into products, fully aware that objects are the category of everything.

complex paths > In the handicraft process, unique products become available, even when they present the same typological and morphological characteristics. A piece in a homogeneous series is thus differentiated from the other by a sort of controlled defect, by a greater pressure of the hands on the material, by a deeper cut by a tool, by a different composition of the colour that is used, by the greater or lesser skill or experience of the person that has made it, or by their absolute freedom to introduce new formal or functional intuitions in relation to the mood of the moment. Artefacts of industrial processes, on the other hand, have no defects; the components allowing them to work are electronic circuits, microchips that, with their impulses, can bring life to the objects themselves. This means that the products we use can be the result of activities highly different from one another. On the one hand, there are processes that cannot be separated from the creativity of the individual man in specific settings, from his technical ability, from his manual skill, and from his expertise in using tools, traditional or evolved as they may be; on the other hand, there is the result of processes characterized by complex paths whose phases are entrusted only in part to man, since it is in the technological potential of the machine that their perfection and their performance capacity is expressed. In this last case, these are artefacts produced by highly refined systems, articulated into segments that can be deployed in territories far from the place of conception and design, thanks to remote-controlled machines entrusted with the fabrication phase; the place where the quality checks are scheduled and performed may also differ. But above all, these are processes that precisely through the use of these machines make mass production compatible with introducing the customization of objects.

without interruption > The "Great Atlas of Design", which was published a few years ago, begins with the following words: *From our first action in the morning, when we reach out our hands to turn off our alarms, to the evening when we press the switches near our beds to turn off the lights from our lamps or the pictures from our televisions, we spend our lives immersed in a world of objects. These objects support, house and clothe us. They help us and they also frequently hinder us. This wealth of things has become so well established that it seems natural. Although we are very interested in items of all kinds and spend a great deal of time choosing and rearranging them, we seldom stop to think about the nature of things and the fact that until relatively recently objects were rare, precious and difficult to obtain and preserve. For a very long time, only a privileged few had access to material goods: kings, priests, generals and rich merchants. Everyone else simply had the tools that they needed in order to work and a few other essentials.*

These borrowed words tell a story about design phenomena that is very familiar to us, with its realistic portrayal of a dwelling that is entirely occupied by the objects that have been continually produced throughout the history of design.

Faced with this ceaseless flow of products that is fuelled by its own consumption, we yield to the cyclical, conventional temptation to take a close look at timeless objects which resist planned obsolescence, are not confined to the words of a caption, tend to be named after their functions and are made of truly unambiguous materials. We are in the field of anonymous design, which has already been explored elsewhere. We would prefer to call it by a different name now, but our hands are tied because a sufficiently evocative alternative solution has yet to be proposed in the critical and narrative spheres. An attempt was made recently by the Vitra Design Museum with an exhibition entitled "Hidden Heroes: The Genius of Everyday Things", which was on display at the Science Museum in London until June 2012. The heroism of this collection of objects can be seen in the continual vein of spontaneous and social creativity that has always catered to human needs and discreetly pairs pure intelligence with genuine design of the kind that we highlight.

urging ingenuity > Toys are the objects of desire of children. Like any other part of the material culture, they reflect the social and economic backdrops against which they have gradually developed. Toys are used to imitate the everyday behaviour of adults. Small-scale reproductions are made of objects and groups of items that represent the stereotypes of trades and women's and men's activities inside and outside the home. There are miniature versions of houses and all of the things that take place in the kitchen, the bedroom and the living room. There are miniature recreations of types of work and locations, such as farms, schools, hospitals, fire stations, railway stations and space centres. Toys can play didactic roles or encourage practical skills, with building games such as Meccano, Lego and Playmobil. Mental agility can be promoted by jigsaw puzzles.
The creativity of children can be developed by various kinds of toys, including those featuring characters from weird and wonderful tales, comics and cartoons.
Devices in children's toys can make them active, as is the case with clockwork, manually rechargeable and battery-powered mechanical items, electrical products and objects that utilize electronics and computer technology to interact with users.
A look back at the historical evolution of toys gives an overview of the anthropological, cultural and social system that helped to shape them and the production system which has moved from craftsmanship to industrial manufacturing and seen natural materials being replaced by artificial ones. Manufacturing companies have built up around toys and established themselves as genuine brands with their own distinctive approaches to play and physical qualities. Their output fits the description of industrial products in every way. Specialist professional figures come up with a concept which is then elaborated by designers. The items are manufactured in industrial processes before finally becoming the focus of marketing campaigns that make them accessible all over the world.
The big names include companies like Fisher Price, which established itself on a global scale in the 1970s and 1980s with the help of designers such as Edward Savage. Walt Disney Pictures made its first products back in the late 1930s. Today, Pixar Animation Studios has followed its lead, while American Girls is a prestigious maker of dolls that are in huge demand among American girls and adolescents.
Design is playing an increasingly important role in the world of toys, as demonstrated by the

Kaleidoscope House and all of its parts, which were produced in 2001 by Bozart Toys and conceived by Laurie Simmons and Peter Wheelwright as a dwelling for the new millennium.
The house has coloured, transparent walls. It is designed to offer children a new approach to play and most importantly to introduce artistic products into their lives from an early age.
The furnishings were created by renowned contemporary designers: Jasper Morrison and Ron Arad were behind the items in the living room; Karim Rashid designed the dining room table; and the beds are by Laurie Simmons. The artworks are by Barbara Kruger, Cindy Sherman, Peter Hailey and Allan McCollum. The fashionably dressed parents and children of the Blue—Green family live in the Kaleidoscope House, which is the first part of a collection that could become a city over time.

architectural object > Therefore the "gigantism" of industrial products was strengthened, due to Transfer initially and then to Standardization in the relationships of objects with architectural and urban contexts. Architectural products have imported the forms of technology piloted in technical objects. Their nature increasingly resembles that of a machine with the complexity of an industrial product.
They have progressively dematerialized and at the same time they have displayed communicative and symbolic values that represent their additional value but also reduce their shelf life, in the same way as an industrial product. This means that they become obsolete much more quickly.

strategic value > Greater competition in the global market and the increase in demand from developing countries tend to improve the number of product in absolute terms, the quantity and quality of product types as well as their formal and iconic quality — often decisive in a competitive market . Companies discover that design is an important added value in the product's aesthetics as well as a crucial factor in the innovation of production and products, so much so that design is becoming increasingly vital in the productive system and industrial districts.

at the service of sport > The role of sports design is to interpret and translate the needs of certain activities into products. Since these activities are constantly looking for cutting-edge performances, the products require continuous, highly advanced experimentation. The contribution of these products is to increase the performance of the athletes in the sports in which they are used.
At the same time they have to protect them from all sorts of risks.
While I write, Francesco Totti is playing in the World Cup; he is making a comeback after a serious injury. In order that he could take part in the football tournament, Diadora has designed and produced a special shin-guard to protect his left leg injured four months ago. It is an anatomic and ergonomic shin-guard: three carbon fibre layers covered in titanium shield his fibula while inside it is lined in Kevlar.
This is just one example of how sports design involves a whole range of complex, varied aspects of industrial design, making it possible to personalise a product and adapt it to the user's needs by

Things, objects and products

either focusing on a particular design or colour of its formal characteristics or by modelling it on a person's specific anatomical features.

anonymous but industrial >   In short, there are products that we accept as being part of the culture of design: the added value of these products is in their design. They are represented by design because they are "branded." They are recognised as symbols of academic experimentation, objects of desire thanks to their emblematic value compared to specific figurative cultures, high-tech products or new productive processes. Next to these products there are many, widely distributed, industrial products: a countless number of "anonymous" industrial products destined for mass consumption and sold either by big distribution chains or through alternative channels working in parallel to the official ones.
More than their "academic" counterparts, these are the products that become everyday items: they bear witness to how big numbers are the end product of industrial production. They represent the "material culture" of our industrial society, post-industrial society and, little by little, of today's globalised society.
These products are produced based on an atopical concept of what are useful objects: they are indifferent to context which, on the contrary, they themselves tend to determine, becoming, from time to time, the paradigmatic representation of an economy, of specific individual or social behaviour, of the widespread taste of mass culture.
These products allow us to reconstruct the contradictions of the debate on industrial production. They allow us to understand how the relationship between design and mass production has gradually evolved based on economic, political and cultural considerations as well as being directly influenced by social models that they have involved or represented: the rationalistic, functionalist Fordist approach associated with social models that used widespread consumption to support the establishment of a more democratic society; the approach programmed to ensure that mass-produced products wouldn't last, a way to increase the size of our affluent society, but which actually creates an tendency towards continuous consumption; the concept of radical-pop culture that by criticising the power of homologation and conditioning exerted by serial products, stresses the value and independence of the individual; or the more recent trends that have stemmed from reflections on globalisation, on the elimination of the distance that regulates the flow of products in space and time in which differences between culture, traditions, customs and systems of social relationship have been abolished by eliminating space and time in communications and information.

between objects and contexts >   The production of the Thonet chairs after the discovery of how to curve wood is a perfect example of the development of techniques and building materials and the deep-rooted renovation of architecture at the end of the 19th century. Crystal Palace is an example of such a development since its prefabricated elements of alloys and glass made its construction possible.

The four examples of chairs and armchairs by Mackintosh belong instead to the artistic revolutions of Art Nouveau, as do the smaller artefacts by Henry van de Velde (crockery, cutlery, etc.) or the handles, columns, everyday or urban objects by Victor Horta who created a continuum by using an uninterrupted line. All the years of Rationalism or abstract-figurative "ism" represents the history of transversality, of contamination between artefacts destined for everyday use in the home or workplace and the architectural and urban context. The De Stijl experience is noticeable in the small blue-red-yellow chairs by Gerrit Ritveld as well as in the Schroeder House in Utrecht or the paintings by Piet Mondrian. The early, wooden chairs by Marcel Breuer have elements of contamination with neoplastic doctrine. The 1925 tubular metal armchairs and chairs by Marcel Breuer and the 1926 and 1929 chairs by Mies exploit the characteristics of these new alloys and determine the figurative shape and geometry of space expressed by Mies in the Pavilion in Barcelona. Le Corbusier's adjustable chaise-longue is emblematic of the master's patient research on flexibility. The use of plywood technology by Alvar Aalto and his research on psychological comfort and reactivity or the research by Eero Sarinen on the chemical treatment of wood and the use of powerful pressure moulding machines to make "organic" shapes are all linked to the basic ideas of the organic movement. So Arne Jacobsen's stackable chairs plasticize wood and create neo-expressionist shapes.
However, the relationship between objects and contexts has gradually changed. Initially, it was a relationship we could call *transfer* and then it became a relationship we could call *homologation*.

sincerity and ambiguity > Nowadays, materials are used according to their expressive style, using what we could call symmetrical procedures. In fact, there's a new way of interpreting the communicative contents of a material that is developing alongside the traditional one: undoubtedly, it's an interesting phenomenon to study. In a nutshell we could call it: the "sincerity" and "ambiguity" of materials.
Obviously this approach has benefited from the continuous innovation produced by research on materials. This innovation involves: ultraperforming materials that are more resistant, lightweight, long-lasting and flexible; multidimensional materials that are physically three dimensional and create a substance with completely new characteristics and appearance; repurposed or recombinant materials, recycled, remixed and transformed materials, almost surrogates used in different ways and for different purposes, or just recombined; intelligent, transformational or interfacial materials that interact with the outside world and adapt from time to time or that interact with our senses based on certain assumptions or that are influenced by biological systems, or custom-designed in their DNA at nanometric level thanks to the evolution of the science of materials similar to the one introduced by the digitalisation of information.
Materials have added value as far as performance is concerned, but also a greater ability to interact with sensorial stimuli. This creates new styles that affect man's behaviour, the shape of objects and a different and more complex way to relate to the latter through his senses.

Things, objects and products

technological innovation > Techne uses tools and procedures to make production easier and better.
The artefact is the product, the prosthesis, that will satisfy some of man's needs, be they functional or not, to improve his way of living in today's environment.
The process that transforms a need into a prosthesis is based on man's inventiveness and creativity.
Man's inventiveness helps to design a product that satisfies his need, defining its performance criteria. With techne, instead, creativity organizes and controls all the materials necessary to achieve the pre-established goal and the work methods and tools needed to produce the product/artefact.
The relationship between techne and artefact is the visible representation of evolution.
In practice, technological innovation affects the evolutionary dynamics of technical objects which up to now have been, and are, material products which, like extensions of the human body, carry out specific functions more elaborately and efficiently. However, these functions are already part of man's sensorial faculties.
Nowadays, technological innovation affects the development of immaterial products as well as technical objects, in other words material products present and visible in the artificial landscape of our everyday lives, our domestic space, workplace or urban space where we move and act.
The actions and processes activated by technological innovation affect our bodies or the biological mutations of animated beings since they directly affect our bodies by producing biological variations capable of introducing into these artificial products procedures that are foreign to our sensorial faculties, giving it new empowerment.
New technologies work on biological processes, creating genetically modified organisms. In other words, they allow the concept of artificial and industrial products to be extended to foodstuffs: once the fruit of a natural growth cycle, the colour and size of tomatoes, courgettes and other foods are designed to fully satisfy market demand.
New technologies make artificial products "intelligent", making it possible for them to interact and dialogue with man, understand his moods, listen to him from a distance, increase his wellbeing or act as a go-between.
New technologies not only allow artificial organs to project our bodies outwards, but also to directly integrate new functions into our bodies, i.e. to see in the dark, to speak and listen underwater or at a distance, to see beyond reality. For man and his relationship with science and new technological tools this is not new, but never before has the potential of technological innovation pressed us to reach beyond the limits of what is possible.

hybrid products > In practice we are faced with an clear alternative. On the one hand, exclusive products whose semantic value lies in their uniqueness, and this depends on the artisanal nature of their production. They are recognised as being "ethnic" products because they tell the story and illustrate the culture of a race or an ethnic group. On the other, mass-produced objects which aim at some sort of perfection, be it to do with performance or morphology, and where exclusivity depends on the innovative role of technology. We call these objects, "technical" products. But due to the current merger between these two types of production methods, this alternative is almost non-existent and the consumer is faced with a truly hybrid phenomenon that applies to all the artefacts of "mixed" objects, in other words neither "ethnic" nor "technical". Instead, objects are "ethnic/ethnic" only when they actually come from an artisanal tradition, "ethnic/technical" or technical/ethnic' when that tradition is copied by machines whose real potential would produce completely different results and "technical/technical" when the products exploit the full potential of technologically sophisticated production methods. This dynamic is semantically so strong that it seems possible to say that this hybrid phenomenon is the dominant characteristic of the abovementioned *global style*.

design museums > Turning a technical object into a fetish to be preserved in design museums arouses a reflection.
The mechanical, electromechanical and electronic parts of a technical object represent its soul and make possible the relationship with people through the sound, the image or the movement produced.
Parts are the essence of the object.
When they break down, the object dies. What is the sense of a design museum? Is there any sense in preserving and showing manufactured products and technical objects that have lost their original use, turning them into fetish?
In design museums we may find silent radios, blind cameras, switched off monitors and still typing machines, all arranged in shelves that make us feel as we were visiting a junk dealer instead of a museum, especially when not too much time has passed since the object was still in use.
In my opinion, science and technique museums should have the function to document and record industrial products in their evolution system. I do not see any sense in showing as a fetish Lettera 22 in a design museum among chairs, tables, televisions and so on. I think it is more useful to show how the technical evolution of an artefact, built for the first time in 1753 by Friedrich von Knauss in Vienna, slowly led to the first industrial typing machine built by Remington in 1873, or how the first typing machine with visible letters (Underwood no.1, 1896) led to the first electrical typing machine (Cahill Writing Machine Co., 1901), and so from the first electronic typing machine to the present generation of intelligent machines.
In my opinion, the formal and aesthetic content of technical objects can be highlighted in museums of contemporary art by including them, from time to time, in specific and thematic exhibitions or studies on specific languages and styles.

upgrading > With globalisation the flow of products and money has become a circular one and the big brands are creating the society's image turning individuals into consumers. Some products are particularly representative of this condition. Coca-Cola, jeans, and the latest generation game consoles, from Nintendo's GameCube to Sony's PlayStation. Coca-Cola, with its material product and immaterial brand — the unmistakable red and white label — has penetrated over 200 countries and has a market as big as the planet itself. As Jean Baudrillard maintains, it is the emblem of the Zero Degree soft-drink; it is a fetish item in a universal, untranslatable advertising language with no definition other than its brand. The product was the emblem of a conscious communication project right from its beginning when the name Coca-Cola was considered fundamental: a brand utterable in the same way in almost every language in the world. More money was invested in its advertising than ever before-in 1908, only 12 years after it was first marketed in 1886 in the Atlanta drugstore of John Stith Pemberton, Coca-Cola was already occupying 2.5 million square metres of advertising space and its logo was in more than 10,000 shop windows. Jeans symbolise a crossing over of limitations of gender and social standing. Not representative of any particular culture, they represent all of them, without any ideological message if not that of the opposition to differentiation — except when they contradict that same generating principle, to become a customised fashion item, with different cuts, shades of blue, stitching and closures, and logo. But this continuing quest for novelty produces nothing but jeans, with their shape, texture and studs. Their durable cotton denim is always assured to wear and age well, and to improve with every washing. Latest-generation consoles like the Sony PlayStation are representative of a way of playing that is the same throughout the world. In the past other toys had already had the capacity for market penetration in many countries around the world, such as the Barbie Doll and the Playmobil. Nevertheless, these were toys that, while homogenising desires and rules of the game, still made it possible to safeguard children's imagination and their relationship with the specifics of tradition.

Finally, the product's use and disposal. The object tends towards increased performance in qualitative and quantitative terms. In highly technological objects only a few of its many uses will be accessible to the ordinary user; there are no replacement pieces for its simpler parts as maintenance is foreseen only for its more complex parts, or else it is meant to be "up-graded", that is, substituted with a new and more evolved version. This easy disposability is creating a major ecological problem and underlines the excessive consumption of resources that our planet will not be able to sustain if not by means of a new approach to manufacturing, conservation of materials, limitation of energy consumption, recycling of discarded products and the use of eco-compatible methods. This is a scenario that leads me to the belief that a real cultural revolution is necessary, one that concerns globalisation in all its aspects (economic, cultural, political and institutional), not as a factor of imbalance and inequality but as one of agreement and development that respects the rights and opportunities of man in the contemporary setting in harmony with his history and culture. And my hope is that the world will decide to think about the future with an awareness of the limits of what is possible.

§
[2]
p.66

Ethics and aesthetics

the toolbox
from Il linguaggio di una lezione, Lectures 3, Roma 2014

doubt
from Lezioni, letture e parole, Lectures 2, Roma 2014

training
from Il design è ubiquo, Lectures 1, Roma 2014

the agon of a smart culture
from L'intelligenza desiderabile, DIID 58|2014

homo faber
from Dispute vere e presunte, DIID 57|2014

performance
from Mutazioni, DIID 44|2010

into chaos
from More than 100 designers, DIID 33–34–35–36|2007

how, what, who, when, why
from Difference&Design, DIID 24–25|2006

light or strong
from Difference&Design, DIID 24–25|2006

the reasons
from Mass design o il potere dell'oggetto anonimo, DIID 15|2005

the aesthetics of high tech
from High Technology, DIID 9|2004

coexistence of systems
from Ethnic & technical, DIID 6|2003

ethically-minded
from Global design ed ethics plus, DIID 3–4|2003

the dominant taste
from L'arte della riduzione, DIID 2|2002,

For many, especially in the design disciplines, ethics and aesthetics are two sides of the same coin. Ethics in the broad sense deals with forms of human behaviour to which to attribute political and moral values. Aesthetics commonly regards the search for beauty, the sensitive beauty of our surrounding environment. Ethics is an ideological condition, while aesthetics is an experience. Ethics allows us to sustain aesthetics and aesthetics is necessary to conduct ethics in its operation. There should be no design at all that does not contemplate, in its phases of its conception and material configuration, in its life cycle, the theoretical and practical combination of the terms "ethics" and "aesthetics." It is an aspiration that probably appears ideologically complete, but is less and less present in design practice.

Ethics and aesthetics

the toolbox > We are well aware that, in the design field, "images are words," and as such possess an absolute linguistic autonomy; however, albeit with respect for their eloquent communicative foundations, it may be hoped that the use made of them will complement the responsibility of a complete, authorial, and non-fragmented verbal discourse. Maurizio Ferraris says this clearly, giving three absolute values to the meaning of language. He asserts that language must always be the result of a thought (there is no true language without true thought, because thought is the mirror of the mind); that language possesses a pragmatic-operative nature (that is to say, understanding that it can be used under certain conditions, beyond the mere descriptive functions); that, lastly, not everything can be language, and that therefore language cannot do everything, also because "the crisis of a civilization is presented as a crisis of language. Language may be the way of truth, but it is also the way of lies, and of loss of meaning."
According to a well-known metaphor, language is a kind of "tool box"; each of these tools serves a different purpose. For example, as British philosopher John Austin has shown, "(language also serves, ed.) to do things, to build objects of a certain type [...] it is not limited (only, ed.) to describing and communicating, but builds objects." Austin's words are more than enough to establish how necessary it is to grasp the complexities we are dealing with (from the scientific and methodological standpoints) when, regardless of content, we examine a lesson's authority with respect to the language used.
Ugo Volli states that "man is the animal that speaks; he lives in language. This statement underlies all philosophy, from Plato to Heidegger. Today, we have not changed our minds from that initial intuition, but (we may place, ed.) language in a broader and relatively new category-that of communication." This means we cannot forget that "language is communication" (and that "the medium is the message," as Marshall McLuhan wrote in 1964, observing the mass media in relation to modern society):"despite this, it bears remembering that every communicative action is always an act of dialogue, and a dialogue always entails an ethic, because it presumes more than one person exercising it through democratically shared conventions ".

doubt > These observations raise questions on the meaning to be given to our initiatives in the area of the culture of design and the role these initiatives have inside and outside our "intellectual circles." In short, it is always doubt that moves all. For Descartes, it is the origin of wisdom. The origin of questions-of all kinds.
There is the title of a book from a few years ago that is suited to our case, written on the occasion of an exhibition on Ettore Sottsass, and it is he himself, as one learns from reading of text, who came up with it: *Vorrei sapere perché* ("I would like to know why"), from 2007–8. To his interviewer, Sottsass declares that he always has "a position of doubt about things. I never have a good idea why they happen." But they do happen. Some pages later, he offers his point of view, which is relevant to our case, on the term "avant-garde." He is convinced that this is a beastly word:"I have it in for the word avant-garde. I am not interested in the avant-garde when it is defined as such. I am not

interested that someone thinks they are on the cutting edge. When we were making Memphis, they would ask us what revolution we wanted to make. They treated us as the avant-garde, in fact [...] but we were not the avant-garde. We were poor wretches who thought we could see what goes on when designing with new things, with new materials." I consider the tendency for the new that Sottsass evokes as the necessary aspiration for the layman's knowledge of the things that surround us-including their fragility. In this setting, the design should always be waiting to understand its role, and ready to change the operating tools because everything changes with fewer constraints and greater empiricism. We then have the possibility, as scholars and teachers, to understand that design culture's demands for emancipation are now formed spontaneously, unexpectedly, and even improperly. And this is a reality that cannot be underestimated. It must not be.

training > Therefore, design is expressed by convention in the ordinary dimension of the everyday, and therefore is not only recognizable in the objects accompanying a part of contemporary society, but is, rather, expressed as diffuse design applied to stable goods, conceived for all productive and entrepreneurial sectors; for a scenario that clearly and tacitly expresses the rhetoric of the design needs of advanced industrialized societies .
In its traditional task of giving shape to products, design renews its disciplinary foundations, comprising within its sphere of reference all the complexity of industrial design. Training, then, is lifelong, and requires a learning system able to transfer advanced knowledge, but also the ability to understand the connections between supply and demand; that makes it possible to know the instruments for managing complexity; that can provide the methods for analyzing demand and translating it into innovative articles of use in relation to performance requirements in addition to the production technologies.

the agon of a smart culture > This is why those who deal with design, now and in the future, must grasp that the "intelligence introduced into things" requires, before anything else, a personal, theoretical, methodological, and objective awareness of the role of the "intelligence in things." They must be made to understand that when new scientific knowledge produces new technologies-whose applications modify and increase the characteristics of the materials, or even invent new ones-the paradigms of the consolidated design models, and above all a design's reasons in being transformed into a product, change. It is unthinkable that one need merely switch the head and the feet of some mute object and then give them life with "wireless devices" in order to access the *agon* of a "Smart culture." A sterilized recovery of established types, integrated solely with new technologies in order to "enrich" products and the product systems with contemporary performance features, is also improper. Probably, the only true intelligence that does not betray a sort of "smart philosophy" nowadays is that of the materials that also — given the commitment of designers/experimenters and of progressive companies — play the basic role of an accredited smart innovation. To provide an intentionally naive example in the field of textures, the much maligned Formica was unable to

catch on immediately as a substitute for previous materials without simulating their characteristics, as it did with wood (it would suffice to recall how Roland Barthes distrusted plastic and then went on to exalt it in his memorable *Mythologies*). It took years for Formica and laminates, and plastic in general, to catch on for what they were and not for what they could look like. At that point, once its composition and workability were perfected, Formica became the way to introduce new and "marvellous" colours into homes yet another proof that there is always a "sincere" way to use materials, as well as a "deceptive" way. The former is measured with the autonomy of a new language, and the latter deceives itself.
The aesthetics of high technology produces the achitecture that appears as giant objects: buildings formed by large sinuous or anthropomorphic forms that accommodate spaces for the most diverse uses, without touching the specific activity or dimensional scale or references to historically consolidated types,
The aesthetics of high technology produces artifacts that appear as giant objects: buildings formed by large sinuous or anthropomorphic forms that accommodate spaces for the most diverse uses, without touching the specific activity or dimensional scale or references to historically established types, "Metropolitan Monsters" effect of superficial transfers of the formal and technological features of everyday objects.

homo faber > This dispute, not at all apparent, on the figure of the designer transformed in consideration of the economic and political changes of social consumption, is the direct consequence of a value-based approach, in which the cultural orientations of open source, of open design, of social design, of co-design, feed on their own, restoring a public model of modern design such as to surpass that romantic and individualist dimension that can communicate the activity of a single person. A design given new foundations in this way acts as a bridge between the exercise of personal talents, collective talents, and new paradigms of utility and beauty that are understood today as accessible to all from every standpoint and under every status, because they are culturally sustained by social utility experiences. There is a claim to play an intellectual leading role-role that gives to those who do something responsibility for the full result of their actions. Therefore, if we wish to find a consistent linkage between these different positions, we must perhaps wait for the principles of the fact at hand-whether with manual or computerized equipment-to encourage the re-emergence of a modern and up-to-date *homo faber*, the man that, in the hypothesis of philosopher/mechanic Matthew Crawford, rediscovers labour as "medicine for the soul" in a moment when professional careers are crumbling, skills are overturned, and aptitudes old and new are again up for grabs. This was discussed even earlier by Richard Sennett, who raised the need for a "craftsman" who must put traditional and current knowledge into play within a modernity that requires a mobility of thought, and an unprecedented tendency for design. The authors of this change must be individuals capable of freeing themselves through their own professional evolution in a society in which skills tend to become quite quickly obsolete.

performance > Had a chance to visit the new MAXXI, designed by architect Zaha Hadid, for the inauguration that saw more than 20,000 visitors come together in an event that has become even more important than the architecture itself. MACRO, the Modern Art Museum in Rome whose extension was designed by Odile Decq, was also inaugurated at the same time.

It was the opportunity to witness a mutation: two great architectural works dissolving into a performance, transforming themselves into two large, temporary installations.

Architecture whose intrinsic feature is that it is a tectonic work was perceived as the ephemeral setting for a performance, an event that mobilised people and things around art.

In what is considered the Contemporary Art Museum of the 21st century, the Anglo-Iranian artist interprets the flows of visitors in a fluid manner, through ramps and promenades.

Visitors become the protagonists of the space itself and movement eventually fills every perception and every viewpoint, to such an extent that any possibility of orientation in terms of space and time is lost. In what should be the celebration of a new architectural work, open to the city and to the whole world, the architecture itself-made of structures and concrete, walls and roofs, rules and proportions-appears to pass into the background, to present instead the mass of visitors who, as well as filling every space, become interpreters and actors, turning out to be the main aesthetic and environmental element.

In these two different and distant places within the city, the MAXXI and the MACRO, the focus is not on their architecture, but on the complex dynamics that have transformed them into a media sensation. Their two architects Zaha Hadid and Odile Decq are stars in the world of architecture and there has been a flood of temporary events — exhibitions like "Chance encounter on the Tiber", the piece by composer and vocalist Lisa Bielawa, linked to Robert Hammond's urban happening 'chairs', the *Calamita Cosmica*, De Dominicis' skeleton, with its regal nose bidding a sarcastic welcome at the entrance to the MAXXI; together with the many people who have trodden the transparent, high-tech catwalks that cross the dominant red and black colour scheme of the MACRO and the fluid procession that gently crosses the flowing, aerial structures of the MAXXI.

There has been a tide of works, artists, critics, music, video, curious visitors, social events and media; a set of animate and inanimate elements that combine in a great, fascinating performance. And the MACRO and the MAXXI have been a part of this performance, the backdrops and structures of a staging that has spread all round the city.

This is a performance that one witnesses with the emotion that is experienced when viewing a work of art.

Indeed, the MAXXI and the MACRO breathe, live and express themselves in and around this performance, as protagonists in a great installation. Here, two dialectically and physically differentiated qualities of space, use of colour, language and different technologies are compared even though they are apart: two contrasting works of art, which after this great performance go back to simply being buildings.

These new architectural works, although far removed from the great Roman architectural tradition, definitely have a power of attraction and the expressive force to become stable reference points as time passes in the city, irrespective of the dynamics linked to their use.

During my second visit to the MAXXI, on an ordinary and certainly less crowded day, as well as its public dimension, I could feel the connections between the architectural and artistic forms. On this second occasion, the unstable, performance-based nature of the architecture appeared even more evident, made up as it is of spatial setups, multimedia technology and video screenings that help present art that is far-removed from simple figurative representation and requires new technology. Once again the content plays a more central role than its "container", which transforms in a fluid manner to tell a new story every time. It would therefore be virtually unthinkable to interpret the sense of this architecture separately from Gino De Dominicis' 24-metre skeleton lying under the entrance colonnade, or Mario Merz' igloo, the monitors that present the architecture of Carlo Scarpa in the exhibition dedicated to him, or the noise of the drill in Diller Scofidio+Renfro's installation.

In a through-line that links this new Roman architecture with many designs presented in this monographic issue of DIID, architectural design becomes design of an empty space, whose performance-based and theatrical dimensions enhance its features. In the case of Diller&Scofidio's design to convert the High Line (an old aerial railway in Manhattan) from industrial archaeology into a public promenade and park, the architect's actions account for a series of environmental factors that will change its appearance over time. It was designed as transit architecture and does not have a single form, but is made up of many images inspired by its visitors and its vegetation will colour it with the shades of the season, Winter will cover it with snow and its urban decoration will change its perception and perspectives. Using organic materials (nature, plants and even visitors) and non-organic materials (classic building materials like stone, concrete and steel) the emerging 'agri-tectura' once again is an empty space that is a theatre of biodiversity in perennial change.

Art provides spaces with a meaning and sometimes imbues them with moral values.

into chaos > But could Chaos be exactly what links so many designers and so many experiences that represent the expression of design in its current form?

Don't we live in a state of Chaos, generated by the flow of products invading our world?

Isn't the multiplicity of expressive forms of the useable objects which surround us Chaotic, generating widespread visual pollution?

And isn't the overabundance of features present in a single technical object Chaotic, with so many functions that the quantity of uses becomes unmanageable?

The world of designers is chaotic, as they are forced to contend with design in all its facets: from producing material or immaterial product, to designing products for the home or body; from designing for free time, or work or play or mobility, to designing products produced in large quantities or design that is intended for "unique", handcrafted products.

In this Chaos, design, with its players and products, expresses a system of ethical and aesthetic

values which invade the artificial environment, it conditions social behaviours and how we choose and use objects, and it standardises tastes, consumption times and figurative styles.
What might appear to be a wealth certainly causes confusion, resulting from an anthropological mutation that has invaded not only western society, but the entire planet. It's as if a multitude of styles, uprooted from the most diverse geographical and historic contexts, were torn apart and recombined to create a hybrid of many languages.
This is Global Style, the semantic connection in which all the figurative expressions of many of the world's cultural traditions are cancelled out by an approved expressive form to which all the others refer.
E' il Global Style is the semantic connection which expresses a contemporaneity, where the weight of tradition is blurred in space and time in the relationship between culture and production. It is used to analyse those forms of expression which, no longer influenced by the physical context but increasingly transversally, are relevant for their social role and represent new "races". In short, Global Style is that dimension of products which is expressed through the languages that are created by contaminations of styles from the most diverse cultures and traditions on the planet.

how, what, who, when, why > Although consumers appear to be increasingly demanding in their request for personalised products, a contradiction does emerge: they are continually searching for a way to compromise between the "social" need to own an easilyrecognisable product that provides a certain *status* and a desire to buy something and personalise it to show that it belongs to them.
Companies respond to this call for personalised products with Mass Customisation, the process that makes it easy for each client to buy a personalised product because its characteristics makes it "different" albeit "identifiable" with its original series.
To achieve this goal, companies use different methodologies and technical processes.
To assess which part can be more easily personalised and if this can be done directly by the client, let's examine certain issues: *what* in an object can be personalised (e.g. a mobile phone cover); *how* (e.g. with an accessory that changes its appearance; *who* (the user himself or the producer upon request by the client); *when* (as soon as it is purchased and once and for all); *why* (to recognise it or be recognised).
To satisfy these demands the company designs and produces a series of standardised parts so that they can be assembled in different ways: using so-called postponement, personalisation takes place after production.
This method can be broken down into subgroups: modularity through repetition, replacement, adapted as if custom-made, amalgamation, application or freely combined. Every object, even when producing standardised products, can be personalised.
To achieve Mass Customisation, it's very important that the company is able to turn every interaction into a process by using standard methods and the information on each client in its database; this information can then be used for other consumers.

The company will then be able to quickly satisfy the needs of the same client because it knows what the client is looking for. However, if it has to provide the same service for another client, it won't have to reinvent the whole procedure because it can exploit what it already knows. This is the difference between customisation and mass customisation: storing the needs of a single client as well as the procedures involved so that they can be used for third parties when the need arises. One of the most interesting aspects of mass customisation is the chance to codesign the product with a client, trying to reduce what some authors call the "consumer's sacrifice".
The client must necessarily commit this sacrifice when he buys a standardised and not custom-made product, given that it has to satisfy the generic requests theorised for that market segment.

light or strong > in the culture of design, difference and identity both play a strategic and complementary role.
For example, when designing a building, "identity" involves the way in which the work is part of a system (an urban context, an anthropised environment, a geographical system, in short a specific *genius loci*), but it also involves the way in which the rules of the reference system allow the building to relate to the overall environment in which it belongs as "alterity", as an object that is in any case identifiable. Two things go to make up identity: identification and individuation. In identification, the architectural work is based on rules (figuration, building type, construction materials and techniques, constraints in size, norms regarding town-planning and street furniture, etc.) which the building appears to respect and with which it has certain characteristics in common: the latter determine whether or not the work is in line with the reference framework. In individuation, the building depends on its own specific characteristic traits, both within its own grouping as well as vis-à-vis the groupings to which it doesn't belong. These distinctive traits can be 'soft' like the ones that characterise a building, even if they are in keeping with the ones that create a harmonized urban environment. Or they can be 'hard' like the ones that make a building stand out in the urban context (due to its size, function, style, specific architectural importance, etc.).

the reasons > Mass-produced products also allow us to reconstruct (see the article by Mario Morcellini, From Horror Vacui to the Latin Approach to Design) the reason for consumption in mass society as the measure of our longings, even in the absence of need, as the affirmation of the individual as part of a system of relations that envisage the partaking of shared experiences.

the aesthetics of high tech > Inside objects, there is a technology that substantiates their multiple uses and intended functions or the material solidity that makes them last aver time. But high tech aesthetics also contain a superficial technology, when clearly technical solutions are used to conjure up an artefact's complexity, or the elaborate use of new materials hint at a product's innovative content and its inclusion in hypothetical future scenarios.

As if there were a link between an object's radical technical solutions, a sort of muscle-flexing exercise, and the concept of progress and evolution.

On the contrary, new technical solutions make the artefact's functional use and design style lighter, more transparent, increasingly immaterial, ephemeral and rapidly obsolescent.

It's then that we turn our attention to the "hardware" of the technique used to design and produce new artefacts rather than the "software" of *high tech* aesthetics.

We believe that a debate on *high tech* design requires more careful examination because it is the key to understanding how the various artistic genres (architecture, design, painting, videoart, etc.) have dealt with the concept of contemporaneity which has been expressed more often by the redeeming idea of technique and its marvels rather than a dialectic relationship based not on ideology but on ethics, analysis and critique.

High tech aesthetics produces architectural works in which the structural framework exploits complex trussing, guy cables and reticular exposed elements that require complex measures to protect them from fire, or lightweight, transparent floors, off-limits for all those who have dizzy spells, with electromechanical or cable systems that use exposed components made of elegant metallic alloys or shiny materials, but rather difficult to maintain and clean and therefore dusty and worn after a few short months.

High tech aesthetics creates buildings shaped like oversized objects: enormous, lithe or anthropomorphic casings that wrap themselves around areas used far the most diverse purposes, without ever coming to grips with its specific activity, dimensional scale or the references to historically consolidated typologies.

High tech aesthetics use curtain walls to cover building exteriors: their colour and luminosity change depending on the led or liquid crystals that even something simple like a short circuit shuts down and turns to grey like giant, switched-off TV screens.

Like a dog biting its tail, an increase in architectural technologies corresponds to an increase in its ephemeral state, and an increase in its functions corresponds to an increase in its own obsolescence.

Even with regard to the form, function and materials of everyday objects, *high tech* aesthetics is more representative of the symbolic values of the potential of technological innovation rather than the formalisation of the applications and technologies that make the products less ephemeral, useful and suited to the activities they are meant to perform. Technology is used in the miniaturisation of telephones where the numeric dials or displays are difficult, if not impossible, to read if one is slightly short or long-sighted.

It is also used to create the disproportionate number of functions that each technical object is said to possess, without the user being able to access or use them all.

coexistence of systems > Instead, the Ethnic & Technical of the Opening section tries to assess some characteristics of globalised production systems that involve the use of technologies and in particular the tendency to move products from non-western countries to the industrialised west,

giving unmistakably ethnic traits to much of what is commonly used in fashion, interior design, communication and technical objects, in short, in many artefacts used in our homes, workplaces and during our free time.
We believe this topic can be studied based on certain objective characteristics of the current productive system and its technologies, just like all those successful products that need ever bigger markets.
At present, artisanal production systems, mechanised systems and systems with computerised equipment and machines where man plays a minor role all co-exist Artisanal methods create "unique" products even if they have similar typological and morphological characteristics

ethically-minded > What sort of ethics is regulating the artificial products that are beginning to flood our lives? And what sort of aesthetics? An ethics closely related to the quality and quantity of industrial artefacts: material and immaterial products taking over our bodies and our environment-a limitless territory invented by man; products that have surpassed the limits of our perceptual possibilities and which concern an enormous number of categories of merchandise.
Scientific knowledge, technology and production capacities no longer intervene solely on behalf of utility objects and technical products but also include the manipulation of living beings. If we consider, as a characteristic of the industrial product itself, anything that can be patented, we must also include inanimate products, virtual products, pharmaceutical products, plant varieties, microorganisms, genes, animals and even human cells and proteins. The aesthetic expression is a languages that results from the mingling of the styles of the most diverse cultures and traditions on the planet. As the most distant places become more and more accessible, new hybrid languages are produced, born of the mixture of disparate stylistic and notional expressions. Languages that fashion rapidly consumes in order to come up with new ones, ever richer in mixtures and references to the traditions of distant contexts. Universal languages that synthesise and simplify as they proceed by stereotypes towards forms of cultural agreement. Industrial products — utility objects — describe the scenario of our lives; their space and market is the entire planet. In what ethics can we see reflected this ammalgamation of customs, traditions, languages, materials, and "regional" ways of production within the global village? What ethics makes it obligatory to wear Nike shoes? What ethics is behind the choice, between two equal products, of the one with the logo — even if it costs more? What is the ethics behind the circulation of a pharmaceutical product in the first, third and even fourth world? What regulates the use and control of communication products or Internet navigation? What are the values upholding the use of innovative technology for the production of invasive technological objects, over-crammed with more functions than required by the needs and capacities of the end-user? The technologies capable of zeroing the space variable are at the basis of globalisation and subject the variation of time itself to domination. The technical drawing is depersonalised by the universal language of Autocad and the Internet; the drafting of the project and the executive drawings of a product can be seamlessly carried out in the most distant points of

the world and allow for continuous work 24 hours a day employing personnel only in the daytime hours.

the dominant taste > We live daily in the presence of an enormous number of artificial products, objects that give definition to the various scenarios of our domestic lives. They are the technological artefacts with which we interact and which are the icons of our "material culture". They are tools that share a variety and multitude of formal connotations, making up a constellation of products with no stylistic unity but are rather the expression of a multiplicity of languages. Missing is that exploration that creates a dominant "taste", that line of research aimed at formalising one language, dominating or juxtaposed to another that wants equally to state its supremacy.

But it is precisely the co-existence of so many and diverse languages that is the characteristic of the dominant "taste". This apparent richness — surely also a condition for disorientation — is, in fact, the result of an anthropological mutation that has affected not only Western society but the entire planet.

It is as if a myriad of styles, stripped of their geographical and historical roots, had been sewn like so many fragmentary seeds for the propagation of as many new languages. The result is a new International Style a "Global Style". The International Style was the result of the practices of the Modern Movement; of an artistic experience expressed in architecture, painting, sculpture and the applied arts by means of a series of avant-garde movements that brought about the world-wide diffusion of theories, methods and languages indifferent to their contexts: a skyscraper for Baghdad or for Chicago offered equal occasions for stating the same constructive, functional, stylistic and technological principles.

The "Global Style" expresses, instead, the condition of our times in which the quantity of expressive forms no longer coincides with the variety of products produced by cultures and traditions with their own expressive specificity.

This co-presence of numerous styles is a constant throughout the so-called global village and creates, in fact, the condition for a planet-wide acceptance of an intermingling of regional cultures. As for technology and performance, what still prevails in products is the most and best of that which industrially advanced Western societies have been able to express.

But what is significant, on the other hand, is their stylistic intermingling of cultural languages far distant from those of the "old continent", showing clear ethnic traces in their design, colours and materials. It seems that the Euro-centric conception of culture is gradually being supplanted by the traditions and cultures of emerging countries whose potential for growth and role is beginning to prefigure their front-line function and their capacity to impact on the future of scientific and technological development.

In order to orient ourselves among the difficulties in finding the theoretical coordinates and references for the dissemination of this multitude of languages, I think it would be useful to look carefully at one of the many notional expressions and theories upon which the design and production of contemporary artefacts is founded: the approach that takes reduction as its guide. In this sense, Less is More — i.e. the theme to which we dedicate the opening of this issue of the review-is intended as a reference to an axiom emblematic of one of the most significant phases of the events of the last century, by means of which the great transition from the practice of internationalisation to the present one of globalisation was accomplished. Much more than on the complex and variegated panorama of current events, it was the spirit and motivations behind the rationalism of the Modern Movement that provided the roots for Minimalism: one of those phenomena, those fragments of expression within the global system which interests us in this analysis perhaps because it is the most Western, the most ideological, of contemporary styles; and perhaps because it falls in line with a reduction ethic that is in contrast with the exhibitionist logic that animates the majority of material and immaterial artificial products making up the environment in which we live.

The aesthetic of reduction lies not in the simplification of the performance and technological complexity of objects which, on the contrary, are being enriched with new and more numerous uses. Rather it represents the need to organise the rituals of daily human existence, to raise that which is complex to a more profound quality that goes beyond the purely aesthetic — the constant control of reason in the creative process. The aesthetics of reduction means the ability to select the useful from among the superfluous, to understand the most profound *rerum natura*, to balance *esprit de finesse* with *esprit de géométrie*. The theme of the "reduction" and "de-materialisation" of industrial products has been the object of study and experimentation in Italy since the mid–1980s, when the most important contributions led to the development of a design approach based on "doing more with less" — a very different approach from both the rationalist functionalism that developed within European modernist practices as well as from the radical design of the early 1970s, with its extremely radical counterculture approach calling for privation and austerity and the destruction of the object

Useful and useless

lifelong education
from Quello che è utile fare e ciò che è utile dire, Lectures 2, Roma 2014

memory
from Lezioni, letture e parole, Lectures 2, Roma 2014

the crisis of architecture
from Il design è ubiquo, Lectures 1, Roma 2014

an integrable design
from L'intelligenza desiderabile, DIID 58|2014

a new balance
from Dispute vere e presunte, DIID 57|2014

a large farm
from Da Orwell a Disney, DIID 48|2011

interconnections with man
from Ecodesign e nuovi scenari, DIID 41|2009

practices
from More than 100 designers, DIID 33–34–35–36|2007

the designing of food
from La progettazione industriale del cibo, DIID 19|2006

cognitive itineraries
from Sincerità e ambiguità dei materiali, DIID 13|2005

the rare and the exclusive
from Design and Luxury, DIID 8|2004

an octopus
from Tool Toy, DIID 1|2002

Discussing utility and its opposite is a fatal combination in the designer's work: inevitable starting from the genesis until the conclusion of a conceptual path that does not always result in an artefact. And it is a condition that also includes the research done in the design disciplines, especially in the field of design, where the boundary between what is appropriate and what is superfluous is, in many cases, a thin one, due to the inability — and it is a cultural responsibility very present in schools — to establish the priorities of a designer's mandate. Clearly, playing — playfulness as such — must never be lacking, and must not be the only reason for a product, but in this case as well it is necessary to be able to seize the moment and the opportunity, so that there is no useless baggage.

Useful and useless

lifelong education > While it may be held certain that the main task of the university, through the higher education that it delivers and the research it practises, is that of impacting the social value of scientific culture, it is less a foregone conclusion that the university originates activities that place the merits of real content on the same level as purely symbolic content. It is less predictable because it is always a wager that is made, betting on the culture of diversity among the disciplines in order to create useful opportunities for debate. It is a debate capable of encouraging emulation, and that does not mix it among sciences which, by avoiding being confined on the terrain of divvying up educational credits, encourage, with conviction, the knowledge they have, as well as their protective walls. And no credence is to be given to Queen Christina of Sweden and her reflections unearthed printed on a market stall, which, in the form of an aphorism, remarked in the long-ago seventeenth century that "men in schools learn what they later have to forget." Upon closer examination, this sentence is not entirely mistaken. School (the university) educates you, also by giving you a self-criticism that habituates you to forgetting certain notions in order to make room for the new that awaits. It is like saying that we offer specializations as absolute values, but we devote ourselves to knowledge as intellectual syncretism in order to combat even those areas of ignorance suggested by boundaries referred to out of convenience as disciplinary boundaries.

Appearing on the ambiguous plane of ignorance is Zygmunt Bauman with some of his statements that prudently come into use. It ought to be said that although Bauman — from what he successfully described as modernity in the liquid variant as a synonym for uncertainty and transformism — suffers from an excessive "prophecy syndrome," it is useful to report what he wrote in conclusion to his book entitled *Liquid Life*:"ignorance leads to paralysis of the will. One does not know what is in store and has no way to count the risks ". Domination through deliberately cultivated ignorance and uncertainty is more reliable and comes cheaper than rule grounded in a thorough debate of the facts and a protracted effort to agree on the truth of the matter and on the least risky ways to proceed. Political ignorance is self-perpetuating and the rope plaited of ignorance and inaction comes in handy whenever democracy's voice is to be stifled or its hands tied. We need lifelong education to give us a choice. But we need it even more to salvage the conditions that make choice available and within our power."

Bauman manoeuvres on a political plane but it is not a level distant from ours, since that which our work perhaps lacks most, the absence of which makes it weak and ineffective, is the sense of its political depth which would help us comprehend — if we were to seek an initial example — how and when to act in the interest of all, so that the good university might survive.

memory > Our acting in cultural terms is always the result of a dialogue that goes beyond the self-referentiality of knowledge. In vogue during the years of participatory politics, the word "dialogue" then freed itself in the term "debate" when it transferred to our territory. How many times have we used the expression "debate in architecture"? Debate on the city? And then the debate on the environment. In more recent years, the debate shifted to the phenomenology of

artefacts, but compressed into the useless contrast between Global and Local. In appearance, we are a people that debates, and we in academia are frequently the prophets of debate, at times the instigators, and quite often the confused executors among many.
I therefore cannot believe that whenever our knowledge is called into question, the responsibility arises to restore to this demand its natural prestige. It is a matter of the authoritativeness that originates from study, from the value held by the principle of the transmission of knowledge, from interpreting, then, the mission of the teacher and of the researcher as a genuine social and cultural assignment. These are facts that we know, but our memory appears not to offer enough room for the obvious.

the crisis of architecture > The transformations of the urban scenarios are continuously shifting from architecture to design, showing that the changes in our cities are taking place not only in the usual direction of architecture, but also more quickly through the expansion of the set of objects. The landscape of our cities, the currents of visual thought, and technology, but also the aesthetics of our habitat, depend on the evolution of design: in its versatility of product, service, and strategy.
For many militant-style architects, Vittorio Gregotti first among them, the crisis of architecture — in language, in style, in the ability to break free as a tectonic discipline — is measured with the "success" of the design.
Design, Gregotti thinks, by involving numerous areas of knowledge and professions, has caused a sort of disorder of thought, making it possible for a building to be not only like a pair of binoculars, but to be designed as such. Moreover, design, as a discipline "compromised" with marketing and production, has polluted the possible relationship between investments and objectives in the field of the artificial, but above all it is as if it had introduced the time factor and the principle of expiration for the architectural product. Design is therefore one of the fundamental engines of the contemporary city's operation. One need merely consider how, in the processes of repurposing disused spaces, surfaces and volumes, urban centres, to develop themselves, are indulging social stresses through the continued work of design in the guise of "producer of experimentations of technologies and technical devices," also within a hard-to-programme spontaneity that requires the improper use of spaces, thus redefining types re-adapted to the new professions. Conversely, between art and design there is a sort of mutual ubiquity in an absolutely clear version: art and design have, over time, eliminated every kind of barrier, finding spaces for common experiences. Outlining these affinities, however, does not indicate that many design objects are equivalent to artworks and as such are to be displayed (this circumstance relates to other factors), or that that many artworks are shown as design objects (starting from Marcel Duchamp's and Man Ray's *Ready Made* works), but that art, in "conceptual" twentieth-century theory, has proposed solutions that design in some way was sowing in its field.

an integrable design > As is naturally the case, the theoretical and material phenomenology of

Useful and useless

design is constantly traversed by themes and words that catalyze their interests and determine their outcomes. As far as I am concerned, while I am still searching for the remains to offset the debts I have always incurred with the definition of "modernity," I find myself burdened by the term "Smart" that has intervened on the future of our urban and domestic lives, with the objective of competing on everything, and with everyone. As a designer trained in the school of Quaroni, with the spirit of the lessons that teach understanding before doing (fewer than eight of them are enough in certain cases), I believe in diffuse design, that of changes with a low content of proclamations. Some years ago, it was enough to say you were aiming for a design that was integrable into knowledge in order to affirm that smart things could be done (regardless of the scale of measurement used); but today, the attempt to do research on the "artificial intelligence" inside and outside of us appears to be compromised by a lingering feeling of inferiority with regard to advanced technologies. Paradoxically, this feeling takes concrete shape in the word "Smart," a key word that — given the propagandistic use made of it — appears to be the only one that can enable the conceptual containment of the most complex "design processes" that aim to give shape to "change." Perhaps it is because I am fresh from a long reflection on the relationship between the word "smart" and the term "architecture," but although intuiting broad suggestions of this word, I cannot help assessing the insidious trap it sets, derived from its rhetorical abuse. It is insidious above all because it is utterly toothless, to the point of diluting any anxiety over the invisible future we thus avoid looking into as true design explorers. For cities, "Smart" means representing an idea of them at their most usable, connected, and thick with online services (all included in a smartphone). However, this virtuous urban system centred upon the new communications technologies does not appear concerned enough to offer a new model in terms of *forma urbis*. Essentially, there are no adequate hypotheses for the "citizens'" interaction with the so-called networks, formalized in material manufactured items different from what already exists. When shrinking the design's scale, a similar reflection may be made in the area of industrial artefacts. We are already pretty good at cataloguing smart objects (objects with performance features capable of interacting, producing actions upon command, remembering, heating, cooling, exciting our sight, stimulating our taste, sight, and hearing), but we are less so in analyzing the aesthetic and ethical dimension of these Smart objects. Because of this, we are in a highly critical phase in the discipline: we are observing a phenomenon in progress: voluble and incompressible, unstable; but at the same time we are called upon to grasp its impacts on all the profiles of the artificial that, from conception, terminate in the physical and material configuration of "things".

a new balance > Whoever is acquainted with my scientific interests, and whoever is current with some of my more "ideological" positions, knows my uneasiness — at times an instinctive uneasiness — before the word "craftsmanship." I do not accept, and I therefore resist, an entire hagiographic, instrumentalizing, and backward-looking culture that wields the term "craftsmanship" to grasp — remaining in our disciplinary specifics — the origin of design (and thus of "industrial design")

and to interpret the future of designing. True or false as it may be, in the attempt to quantify the craftsmanship hidden or visible in industrialized products, many in our environment have forgotten to critically frequent the landscape of industrial production which, in the meantime, has changed its nature in form and content. My irritation, then, was never over the culture of doing, typically and extraordinarily Italian, but rather regarded a series of prophets, masters, followers of craftsmanship (worse if artistic), who intended to bring much of the "absolute truth" of our work to within the "purity" of the handmade. It is a purity that has never existed. But what does exist — precisely because these itches of thought cannot be scratched — is the will to make these opinions of mine react with the "Makers" phenomenon which appears, within the scope of an increasingly ambiguous debate, to have upset and altered everyone's convictions, considering the evolving relationships between design, production, consumption, and consumers. I wish to say that declaring the Maker strategy represents the model of a new equilibrium in the sustainable relationship between supply and demand in the world of artefacts is improper because the season of the manufacturing industry in its traditional mission has not come to an end. If there is this transformation, it certainly did not impact the so-called productive districts not only in the most advanced countries (Japan, Germany, the United States), so much as those areas in Eastern Europe that welcomed the industrial offshoring of those advanced countries. And then there is still Korea, and on up to the giants of China and India, whose markets are true green pastures for new things. This is to say that cities, homes, offices, public buildings (libraries, schools, universities, hospitals, courts, gymnasiums, soccer pitches and athletics fields) cannot help filling up with artefacts in which to concentrate indispensable smart features (just attend any electronics fair, like IFA in Berlin, to get the idea)
[...] There are high numbers of people who design, in symmetry with those who produce. It is necessary, then, not to fall into the deception that those who make "household" machines to self-produce objects for themselves have found the solution to avoid the flow of goods. Except for rare talents (see the recent example of a car chassis in a box), many run the risk of deceiving themselves they are marketing online, and then find themselves behind a stall.

a large farm > Design for Pets. The title says it all: it is an issue explicitly dedicated to the themes, designs and products that are part of the eternal, complementary relationship between animals and man in a kind of "universal farm" that contains the limits and hopes of a common path and today entrusts design with a delicate, "functional" role that is more ethical than in any other sphere.
Design for a metaphorical farm, in fact for two completely contrasting ones. On one hand there is *Animal Farm*, written by Orwell in 1947, where the animals, tired of their cruel exploitation by man, rebel and take control. However, over time, using the same tools and means as their owners, they imitate them in everything from their behaviour to their appearance. On the other hand there is Walt Disney and his imaginary city-farm, inhabited by men dressed as animals, or vice versa, which are positive, grotesque, fun or braggarts. These legends are part of our culture and imagination that

contain everything we possess or would like to possess.
We shouldn't be amazed therefore if our pets are like humans.
Everything (sometimes too much) is available to them, just like all human beings, set against a backdrop of design culture peppered with collective practices and moral issues that depend on the various emotions that humans project onto them (passion, mania, fear and entertainment), which are fully present in the typical mechanisms of designing the artificial that are features of contemporary design.
In an examination of the various design extensions that can be applied to the animal kingdom: on a physical, domestic and environmental level.
We thus discover that the time, space and objects we dedicate to those who share our home (or our tyrannical friends) represent two design themes. While the first interprets the animal environment through objects that they need to exist (imagined and produced by humans, however), the second stigmatizes the theory of a shared living space for different beings through articles that combine animal ethics and human aesthetics. These ethics are sometimes distorted (through zoos and themes parks) into simulacra of natural habitats whose purpose is entertainment and spectacle. This is also a private spectacle that is seen in the growing phenomenon of 'Fashion for Pets', in which human obsessions with clothing and style are transferred to innocent creatures that become the latest fashion victims.

interconnections with man > I would like to analyse a movement that goes against the trend, as revealed in a recent exhibition: Paris Design in mutation.
The theme of change entails the idea of experimentation, and this obviously also includes the issue of sustainability, but without it holding back far-reaching renewal of products capable of making them the expression of a new present and a visionary future.
It is very interesting to note that public institutions (Délégation aux Arts Plastiques/Bourse Fiacre, Bourse AGORA pour le design, CulturesFrance, VIA-Valorisation de l'Innovation dans l'Ameublement) have turned to designers, including some very young ones, to understand where we are going.
It shows great far-sightedness by the country to ask designers to interpret the changes we are undergoing, giving them the task of proposing products or systems that represent the new world and can help to improve life within the new scenarios.
To revamp the residential culture, Jean-Louis Fréchin investigates the digital world and makes use of its potential while simplifying technical objects: *WaSnake* is a shelf and digital communication device, *WaazAl* is a shelf and audio playback system, *WaDoor* is a door and a screen.
In this way, Jean-Louis Fréchin uses new types of products to reduce the quantity of items in the home and their interconnection with humans. He proposes a domestic environment in which everyday objects interface with the net and home computers. He is an exponent of the digital revolution, which is seen as an opportunity to modify the interaction between products and people,

our relationships with space and time, and finally to give a form and identity to the intangible economy.

François Azambourg takes an almost artisanal approach to pushing the performances of materials to their limit: all of the energy is squeezed out of fabrics, wood, metal, resin, glue and foam. The resulting objects seem like exercises in style, but closer inspection reveals them to be the expression of design guided by the economic utilization of materials. He presents a complete collection of chairs that show how a new approach to the interweaving of materials can shape new objects. Their appearance crudely tells us how they are formed.

In the world of Mathieu Lehanneur, our bodies and the environment live together in a huge number of interrelationships: with temperature, sound and the air. Our very existence and our equilibrium are affected by continual adaptation due to alterations in external parameters. Therefore, his objects cause changes in habitats: *Bel Air* is a system that uses plants to filter the air; *Q* nebulizes the air; *K* absorbs daylight and then emits it in the domestic environment; *dB* is a generator and diffuser of oxygen; *C* is an infrared radiator.

The EDF R&D Design Studio creates products that tell users how much energy is being used during various everyday activities in order to encourage all of us to change our habits: when set up correctly *Coup d'oeil* shows the amount of electricity being consumed, as the colour of its special sensors varies; *Semaforo* and *Orloge???* apply similar principles. Experimentation with this group has led to the creation of other items such as those for the production of energy.

practices > In the first case, without ever abandoning their artistic quest to create unique pieces, designers see their work as a continuous creation of forms intended to please all of our senses, works that are emblematic of a cultivated interpretation of the aesthetic of the artefact.

In this case, designers' experiments aim to reflect the possibility of applying new materials and new manufacturing systems to the artefacts they design, to offer up new types of products, often one-of-a-kind pieces, prototypes that generate new languages which have the power to be 'trendsetters'.

These items are one step into the future, and are often veritable works of art, visible witnesses to contemporary culture. In questo senso are the experiences of Humberto and Fernando whose research is based on using poor materials or industrial scrap to design objects that are unique for their expressive efficiency; Ross Lovegrove whose work is examined to understand how to move into the future; Roberto Capucci, the author of sculptural clothing that are works of art in fabric; Gaetano Pesce an extraordinary experimenter into the use of new materials for 'diversified series' of products, and Essaye Myake whose creations express a continuous technical re-examination and renewal.

Ma iI designer act as the interpreters of people's needs, using their ingenuity to transform new needs into new products while eliminating the superfluous, products that are essential in their form and in the use for which they are intended, designed to minimise waste of the materials used, to simplify the production process, and to provide new opportunities for people's activities.

Particolarmente I believe some of the most noteworthy are Antonio Citterio, who is as refined in his formal minimalism as he is efficient in the wealth of technological solutions he implements; Matali Crasset, who attentively interprets our daily rituals in order to devise new products for our new behaviours; Marc Sadler, Bruce Fifield, Hugo Kogan, Andy Davey and Michele De Lucchi, who apply the most advanced innovations to technical products, including home appliances, diagnostic equipment, lighting and communication products, and even sporting equipment; and Ronan and Erwan Bouroullec whose research is based on flexibility, multiplicity, reversibility, modularity and combinations of elements.

the designing of food > We could say that when French nouvelle cuisine became popular at the end of the seventies, new areas of food design were invented or revamped: these areas touched on sophisticated experimentation, innovation or the scenarios of the new generation of natural foods that involve the merger of fields of learning and traditional expertise with advanced organisational logic, technologies and knowledge. The invasion, perpetrated by chefs into the field of design and designers involving the formal and aesthetic qualities of food, opened up new scenarios around food design, for example, the design of dishes, the design of food and for food. The design of dishes involves studying "equipment": this calls for the combined talent of a chef and a designer. A coordinated, pleasant design of the dish and the food impacts strongly on our visual sensorial experience; it involves our sense of smell and sight and it stimulates our taste buds and ears. This requires multidisciplinary skills and expertise that include ergonomics, chromatology, morphology, geometric figures and, of course, gastronomy. Apart from being able to exploit the shape of food and many different colour combinations to decorate the table, food design involves many new experimental fields. One example is the recent performance of Paolo Barrichella's Food Design Studio during the event, Una luce tutta da ... bere, at the Urban Light Walk Exhibition held during the International Furniture Fair. Using Tonic Soda Campari, they created edible lamps. They took advantage of the properties of quinine chlorhydrate, present in tonic water, which when exposed to black light, filters the UV rays emitted by neon tubes and creates a fascinating visual luminous effect. Without using energy, the liquid becomes a light source, letting the consumer move around the room with his own edible light.
Finally, food design involves that huge range of products used to taste and eat food (glasses, plates, cutlery and dishes), table tools (bottle-openers, nutcrackers) and cooking utensils.

cognitive itineraries > Fifty years ago, in his book Mythologies, (Edition du Seuil, Paris, 1957), Roland Barthes expressed contradictory feelings about traditional materials and objects or the ones created using innovative processes. Because they are contradictory, these two standpoints still represent a paradigmatic interpretation of the relationship between materials and object. The first chapter examines the use of wood or non-traditional materials to make toys, 'Current toys are made of a graceless material, the product of chemistry, not of nature. Many are now moulded from complicated mixtures; the plastic material of which they are made has an appearance at once gross and hygienic, it destroys all the pleasure, the sweetness, the humanity of touch. A sign which fills one with consternation is the gradual disappearance of wood, in spite of its being an ideal material because of its firmness and its softness, and the natural warmth of its touch. Wood removes, from all the forms which it supports, the wounding quality of angles which are too sharp, the chemical coldness of metal. When the child handles it and knocks it, it neither vibrates nor grates, it has a sound at once muffled and sharp. It is a familiar and poetic substance, which does not sever the child from close contact with the tree, the table, the floor. Wood does not wound or break down; it does not shatter, it wears out, it can last a long time, live with the child, alter little by little the relations between the object and the hand. If it dies, it is in dwindling, not in swelling out like those mechanical toys which disappear behind the hernia of a broken spring. Wood makes essential objects, objects for all time. Yet there hardly remain any of these wooden toys from the Vosges, these fretwork farms with their animals, which were only possible ... in the days of the craftsman. Henceforth, toys are chemical in substance and colour; their very material introduces one to a coenaesthesis of use, not pleasure. These toys die in fact very quickly, and once dead, they have no posthumous life for the child.'

Even if the author focuses on toys and not on the materials, he is strongly against plastics and very much in favour of wood.

A little further on, in the chapter entitled "Plastic", he writes:"Despite names recalling Greek shepherds (polystyrene, phenoplast, polyvinyl, polythene), plastic — the products of which have recently been gathered together in an exhibition — is essentially a produced substance ... The public stands in a long queue at the entrance to the hall to see how the magic process par excellence, the transformation of material, takes place. An ideal oblong tubular machine built to reveal the secret of an itinerary effortlessly recovered a mass of greenish crystals from shiny, grooved basins, on one side the blue material and ... on the other the perfect human object and nothing between these two extremes except a short conduit controlled by an employee with a demigod, semi-robot beret ... Plastic is less a substance than the concept of its endless transformation, it is, as its common name indicates, omnipresence made visible. And precisely this makes it a wonderful substance: each time, the wonder lies in a sudden conversion of Nature. Plastic remains entirely permeated by this amazement: it is less an object than the trace of a movement."

In fact, the two apparently contradictory arguments follow two cognitive itineraries of the same journey. In one case, we learn about the history of the material before it became a product and how it was represented with "sincerity." In the other case, the material conceals its origins and the complex alchemy that produced it; a material so "ambiguous" to our sense of touch and sight, and yet so exciting.

So plastic is not vulgar, nor wood, elegant. Wood honestly represents the work of man's hands, the work of an artisan and his tools that cut, shave, smooth and polish. Plastic, instead, is a hermetic and ambiguous representation of the transformation process of a chemical substance and its magic in which man is the robot, the demigod that invents a process and lets this constant process reveal itself even after the object has been produced, as the material changes when it interacts with man, time and the environment.

the rare and the exclusive > The success of an industrial product is measured with its spread into the markets and the consumers' agreement. In fact the more the product demand is high the more pieces are produced and the company turnover and profits are high.
This would make you think that "luxury", an attribute that is linked with "rare" and "exclusive" objects destined to few people, is not suitable with production form based on great numbers such as industrial products are. Nevertheless it is not always like this. The definition of luxury and its representation are strongly conditioned from factors such as globalization and multi-medial communication, marketing strategies, social customs.
Today the symbolic brand value has been replaced with the "rare" and "exclusive" concept based on the uniqueness of the piece and on its value given by a complex of attributes such as quality and the preciosity of materials, the hands competence able to turn products into finished artistic works. The connection between the object and the V.I.P. that possesses it and therefore the emulation of desires become more important than the peculiar values of the object itself: its specific technician, formal and functional contents are replaced by possession as emblematic and social condition of belonging to a group.
The possession of objects and their ostentation represent a status symbol as much as "*griffe*", style and brand product research becomes breathless. The aspiration to possessing objects can spring therefore from a conditioned desire. Desire is conditioned from mass media and advertising which operate on these delicate social, anthropological and mental mechanisms that bias our choices towards brands as much as not only pushing us to purchase a brand product but it make us become quite collectors.
These conditioned aspirations exist also when the object of desire has the highest costs such as purchasing a Ferrari's car, an uncomfortable car with a coarse aesthetical conception, functionally and technologically excessive regarding highway codes and city traffic mobility; anyway Ferrari is the brand winning in the F1 World Championship, and purchasing one of this car is like possessing one of the powerful and inaccessible racing car; and also booking modalities and delivery times give to this product more exclusivity such as an object built on demand to be fitted; it is a sort of exclusivity also given by the excessive costs which are not justified even to pay off all the research and the innovation investments.
But the aspiration to the possession, can gush also from a high culture level desire: this is a conscious desire which make us aware to possess something just for the pleasure of our senses, the pleasure of our body, the pleasure spending our personal time.
In the case of material products, certainly this pleasure has something to do with those characteristics of exclusive aesthetic features that can satisfy our eyes, with uniqueness of materials,

technologies, manufacture, and every ingredients which can satisfy all other senses. Sure it cannot be a mass production object, therefore it is difficult that industrial design, which is an activity comparing with mass production and globalization market, could answer as one specialized operating procedure, or high level handmade workers, or an artist.

It's sure a high level culture the desire to possess an art work, which uniqueness is linked to the artist work.

an octopus > Prevailing in the design of objects is an ambiguity and deception — neither its shape not its materials have an organic relationship with the use for which it is destined.

Thus a polyp conceals a citrus-squeezer, something that deceptively alludes to a feather is, instead, a somewhat less noble toothbrush, and vaguely metaphysical shapes of some unknown material are objects whose scale is even difficult to understand.

In this scenario the object comes to be perceived as a living and affectionate presence. It incorporates emotions that act on each individual and, as a result of those emotions, one wants to possess it. The object becomes an object of desire, something precious, affectionate and reassuring that stimulates the entire human sense apparatus.

This is a production trend that has by now become history in the aesthetic of the ToolToy.

[2]

Hybrid realities

lectio
from Il linguaggio di una lezione, Lectures 3, Roma 2014

cross-pollination
from Guardando oltre, DIID 45|2010

interaction design
from Dipendenze tecnologiche, DIID 39|2009

hybridisation processes
from Hybrid Cultures in Design, DIID 37|2008

design for cinema
from Movie & Design, DIID 28|2007

social nature
from Ethnic & Technical, DIID 6|2003

transfers
from Trasferimenti di senso, DIID 5|2003

nanodesign
from L'arte della riduzione, DIID 2|2002

Sciences tell us that a body that brings together diverse, completely different characteristics, is a hybrid form. Design has thus always been a hybrid form in the broadest sense it can be given. It is hybrid in content, considering the quantity of players, subjects, tools, and objectives assigned to it. It is hybrid in the designing extensions of the artificial to which it is subjected. It is hybrid in its natural postmodern, free, and inclusive inclination. It is hybrid as a set of critical and rhetorical activities able to contain the possible, while enabling anyone to be the author. It is hybrid in the languages and tools used, capable of changing in the attempt to grapple with reality and how reality changes.

lectio > From a book edited by philosopher Maurizio Ferraris, *Domande della filosofia*, I cite these words: "We speak every day, we often speak too much, and at times we have the well-founded impression of speaking without thinking. This seems ill-advised to us, not only for its practical consequences, but also because it harms a sort of an implicit presumption of the language: that behind every word, there is a thought. To make this presumption explicit, to see its range and its limits, we have sought to put three points into focus. The first regards the relationship between thought and language. The second, on the other hand, regards the pragmatic dimension of language, which is to say what is done with language other than thinking about and describing the world. Lastly, the third consists of what, in philosophers' jargon, is called a *caveat*, that is warning to beware of what — again in philosopher's speaking — is referred to as "linguistic holism," the idea that everything is language (however strange someone who truly thought this may appear) or that language can do everything. No, this is not the case. There are things that language cannot do". I have described the meaning, including the formal significance, to be given to a "lesson" as *lectio*, considering the duties the university has include that of promoting the social value of scientific culture in a perspective of a common growth of civil society; and while on other occasions I have dwelt on the metaphor of "designed words" to explain the responsibility we have as teachers in the academy in the choice of each individual term to communicate that knowledge of which we must show not certainties, but if anything the virtues of the doubts originating from practices, from research experiments, from intellectual speculations; in this case, through the suggestions originating from many readings. I prudently reflect on the theme of language and on its instrumental nature to give quality to a lesson.

cross-pollination > The distinctiveness of the different disciplines has long since ceased to exist in the relationships between the Arts, Architecture and Design.
The statutory boundaries of the various operational and expressive fields initially became permeable and then blended into each other, with crossovers occurring to such a great extent that a new framework has been established. Architecture, Design and the Arts draw on each other's experiences and experiments. Everything is hybridized, including languages and the methods used to design and formalize objects.
It is interesting to explore this new scenario in which Design's expressive horizons are broadened to take in artistic performances, objects that evoke user items are among the ways in which Art depicts contemporaneity, and Architecture employs the aesthetics of the Design culture, its theoretical and technical materials, and its production processes.
This new scenario, in which the traditional disciplinary bodies of Art, Architecture and Design have been reshaped into a single, new and complex sphere, is effectively and knowledgeably outlined at the latest International Architecture Biennale in Venice.
Sejima presents a different idea at the Architecture Biennale in Venice.
Her intelligent, expert vision of the world involves a concept in which Architecture, Design and Art

all mutually influence each other.

The story that she tells takes shape all the way along the route, with a metaphysical portrayal of Architecture: there is more of a focus on the development of work than its formalization. This is the most effective way of tracking down the building blocks of hybridization.

As visitors wander around the installations of the 46 guests, they follow a sort of peripatetic route that derives from Sejima's outlook and is examined in greater deal in the Thinking about... feature.

Between the spirit and theoretical material expressed by Sejima in "her" Architecture Biennale and the topic of this issue of DIID, there is common ground being explored: the awareness that planning artificial elements now involves a new concept in which architecture is the location for the settlement of aesthetic and ethical values that are shared with the Arts and Design. It is this common thread that guides us through the show.

Although the exhibitors all have their own outlooks, it is plain to see a loss of interest in the typological classifications of buildings and a move towards a more human approach; a loss of interest in monumental constructions and a move towards a light, instable, interactive and modifiable offering that ends up blending with the performance.

Architecture that is designed to stimulate all of the human senses is presented. Just as light and colours appeal to the sight in works of Art and touch comes into play with materials, all of the senses are involved as things move on from the traditional concept of definitively organized functions in favour of a cognitive relationship with the work, its utilization, its perception and the technologies for interacting with it.

There is a new relationship between people and the work, and between the work and its surroundings, whether they are metropolitan or urban settings, or natural landscapes. People take on a central role that has never been so important. The work interacts with onlookers: it becomes "light" due to the transparent features that cancel out its materiality; "impalpable" due to the light that shapes its forms in a continual transformation process; and "dynamic" due to the lack of Euclidean references. In the work there is also a "character" that expresses the genius loci of the place but at the same time makes it an external element, as if it had only been put in the location temporarily or as if it were part of an artistic performance rather than an object. In short, there is initially a little disorientation caused by the work because the familiar patterns cannot be identified, but then you find your bearings and comprehend its status as a new expression of contemporaneity. All of the innovative aspects of science and technology — tangibly and intangibly — shine through to give a new vision of the world in which Architecture, Art and Design merge to create items for playful and fruitful use.

Sejima fascinates visitors and gets them involved with a collection of experiments that all look towards the future. Instead of singing their own praises, they show curiosity about things that we do not yet know, but that we hope will meet our expectations.

In the proposals that have been selected, environmental issues are considered but keep a low profile, unlike in the tradition and culture of militant green campaigners. Sejima believes that the

solutions for fresh sustainability can be found in scientific research, the resulting applications and new technology. She takes a secular approach that acknowledges the supremacy of humans and is in contrast with the ideological standpoint of those who are hostile to new technology. I would call the latter outlook "misanthropic", because it works against humans and denies them the new opportunities provided by innovation.
Once again Sejima places people — rather than the object — at the heart of the matter. The first to understand and appreciate this are children, who are not conditioned by overtones and display wonderful naivety in their senses and their emotions.

interaction design > Gestural interfaces, which allow users to surf among audio and video contents, to download images, texts, movies and music, through hand gestures used as a command on a display, or even only by the means of gestures, keeping at a distance from the display, are possible thanks to the combination of Bluetooth technology and digital camcorders or ultrasound and infrared sensors which include *Body Language*. When in *Minority Report* Tom Cruise looked at a display and manipulated images and maps, browsing virtual pages thanks to the movement of the hands, it seemed to be the representation of a distant world. Today it is the present, or rather it is already the "present perfect". The same technology will drive our choices in shopping centres, will answer our questions, and will give us information in public bureaux, at the train station and hotel receptions desks. It will affect and change our behaviour and social relations.
Those applications are ruled by *Interaction Design*, a field of design researching the cognitive approach of interactive systems, whose theoretical grounds were laid down during the '80s. In those years Bill Moggridge (IDEO co-founder) and Bill Verplank (*interaction designer and human-factor engineer*). first coined and used the term "Interaction Design", overcoming what in the past was known as "user-interface design", thus ascribing to the new discipline of user-interface design. the functions that make Design a discipline integrated to the whole process of product design and development. Later on, at the beginning of the '90s, *Interaction Design* established itself as a special field of design for operation and research. In 1991, David Kelley, Mike Nuttel and Bill Moggridge founded IDEO, a "lab" which soon became a worldwide famous product and design firm, with a user-centred design approach. An approach that can provide design services for the applications of the interaction between human cognitive system and the tools used to process information: in this way, products suitable to the most different uses, from work to education and entertainment, can be created. Research is therefore focused not just on the improvement in the functionality of technological products, but on anything that can meet user's needs in order to improve his life. The idea of user-friendly technology started gaining ground.

hybridisation processes > We know that the Hybrid (from the Latin *hybrida*) is an individual generated by the crossing of two or more elements that differ from each other in a number of characteristics.

In the biological sciences, hybridisation is the crossing of two animals or plants with the aim of modifying some of their characteristics, having others emerge or creating new varieties.
A Hybrid, therefore, always marks the passage from a given form to another that "contaminates" the former.
Ovid left us a text (*Metamorphoses, Book I*) that describes this passage in poetic form.
But this passage also involves the action of *homo faber* who, beyond his imagination, through the factual practice of techniques, is always generating new artefacts.
The theme of transformation, in its positive and negative aspects, runs through the history of humanity. Without transformation, there would be no development.
Design therefore finds itself governing a process of hybridisation that would otherwise be spontaneous, but that inevitably finds itself having to recover its own disciplinary statute.
There, where historically its role was to interpret man's needs to improve his living conditions, using the manufacturing materials and techniques available in specific eras and contexts, to produce artefacts in continuity with the material culture of communities and social groups, today Design has to face up to a moment of change that does not only depend on the potentialities of technological innovation, but also on many other factors: from the contamination of the various local cultures as a result of the use of networks of IT communication and with the increased mobility that is available to the whole population of the planet; from the shrinking of space and time in a globalised society in which contexts are rendered changeable by a widespread culture that hybridises everything in a continuous process of production and reproduction of artefacts: from the change in our desires and behaviour: in eating, in dressing, in the range of artefacts intended for domestic use, at work, studying, during free time.
In effect, today, technical objects, objects we use, that is, the constitutive elements of our material culture, are the result of contaminations that give rise to different meanings of hybridisation.
The typological Hybrid has to do with the complex system of contaminations that have an impact on the intrinsic properties of a product, when, overcoming the whole historical consolidation of the characters that have settled on it, it introduces, starting from the original model, new uses that alter it until it is transformed into a new artefact. But this also involves the simplest contamination of the characteristics of two or more typologies of product that generate a new typology through metamorphosis.
The functional Hybrid has to do with the various forms of use of the product itself or its typology, such as the transfer of typologies of product from the realities relating to the advanced industrial societies to emerging or developing societies and vice versa.
This is the case with products, that even as they maintain their constitutive characteristics, are used differently to how they were meant to be used.
The technological Hybrid has to do with the system of contaminations that are applied to a product to respond to new needs. This involves new products that are born from the need to improve man's living conditions, and that can be planned and produced with new uses, and with the application of

new scientific and technological knowledge. This is the case in forms of hybridisation that use the technological transfer from one field of application to another for new and unimaginable products that represent all the potential of Design: machines whose use we cannot imagine, interactive objects, materials that through the use of nanotechnologies can be applied even to infinitesimally small objects.

The dimensional Hybrid has to do with the alteration of the scale of reference of the object both in relation to its use and in relation to the methods referring to both its planning and manufacture. It has to do with the transfer of the formal characteristics of a small object to a large one, to the extent that household objects appear to form miniature architecture, and architecture appears to be gigantic household objects.

But it also has to do with the contamination of digital technology in the processes of manufacturing the product that, through the use of numerically controlled equipment, should no longer have to be concerned with tolerances, or controllable errors, insofar as there are no longer any possible errors, not even in the dimensions of the infinitely small.

The linguistic Hybrid has to do with the various expressive and stylistic forms of Design products, and it is perhaps the most interesting aspect to study. Partly because the communicative content of Design products is the foundation of the material culture of an era and a context, but also because Design objects often include in their aesthetics the characteristics that are present in the performance of the more consolidated arts, placing themselves in spaces that cross these.

"...the goddess ... uttered oracular speech:'Leave the temple and with veiled heads and loosened clothes throw behind you the bones of your great mother!'

For a long time they stand there, dumbfounded ...

Then Prometheus's son comforted Epimetheus's daughter with quiet words:'Either this idea is wrong or, since oracles are godly and never urge evil, our great mother must be the earth: I think the bones she spoke about are stones in the body of the earth. It is these we are told to throw behind us'.

...obeying, they threw the stones needed behind them.

The stones, and who would believe it if it were not for ancient tradition, began to lose their rigidity and hardness, and after a while softened, and once softened acquired new form. Then after growing, and ripening in nature, a certain likeness to a human shape could be vaguely seen, like marble statues at first inexact and roughly carved.

And quickly, through the power of the gods, stones the man threw took on the shapes of men, and women were remade from those thrown by woman."

Ovid (43–17BC) *Metamorphoses*, Book I

design for cinema > Spectacular movies as an expressive outlet for design: this is the premise for our musings over the relationship between Movie and Design in this issue of Disegno Industriale. The cinema as a tool to create narrative, but also as an art form behind special effects, stage and costume design, multimedia communications and visual design. These are just some of the most dynamic aspects of the movie industry, a sector in which Italy has led the field internationally thanks to the splendid traditions and expertise of craftsmen and Oscar-winners like Danilo Donati, Dante Ferretti and Carlo Rambaldi. If we look at international production, reviewing masterpieces by Stanley Kubrick helps us understand how innovation in the movie industry is not just a question of style, but also of cinematographic technique. For example, the sets and techniques experimented and used to produce 2001. A Space Odyssey show how people imagined the future forty years ago and how design was part and parcel of the construction of sets and atmospheres which even today still look feasible. To talk about movies and design today means again focusing (even after the advent of the computer age and new multimedia tools on how these two fields can be combined to help the user/spectator discover the imagination of designers and directors (Saul Bass, Tim Burton) and ho they portray and inspire feelings through movement, reestablishing the right empathy between objects and people. It's no accident that these are the notes used by the director Peter Greenaway to interpret "Blow the trumpets! Overture−2000 years of Italian creativity" in the newly-opened Museum of Design at the Milan Triennale, an installation that welcomes and accompanies the visitors through an exhibit design in which multiscreen projections show reruns of old Italian masterpieces. The objects "in front of the fluid movie wall" seem to be "part" of the scenes of the feature films in the background; they tell the story of the "obsessions" of design (Italo Rota), or perhaps we should say of some of its more important "categories" or "scenarios" (Andrea Branzi): The Light of the Sky designed by Antonio Capuano, Supercomfort by Pappi Corsicato, Dynamicity by Davide Ferrario, Stackable Democracy by Daniele Lucchetti, Animist Theatre by Mario Martone, The Simple Great by Ermanno Olmi and The bourgeoisie and the sacredness of luxury by Silvio Soldini.

social nature >   a careful analysis of the trend towards homologation of all the products we use in our daily life, both publicly and privately, in both the southern and northern hemisphere. In the so-called Global Village, this contemporary characteristic tends to eliminate the peculiarities of material cultures and traditions which, in practice, belong to specific territorial and geographic contexts and which, vice versa, promote the worldwide distribution of similar objects and technical products. This trend has implications of a social and economic, political and moral, aesthetic and cultural nature. It affects the development of technological innovation and production systems. It is a complex subject and the magazine has focused on its own field of competence, namely the functional performance, technical characteristics and formal contents of all material and immaterial products and product systems that constitute our artificial environment.

transfers > Today, we can raise environmental awareness producing an impact on the management of the complete life cycle of an industrial product by saving energy and materials in the production process, making assembling and dissembling operations easier in the production and dismissal of products, reusing and recycling materials and components of dismissed objects to

manufacture new products.
Environmentally friendly products and, in general, a sustainable Environment are crucial aspects in domestic and international policies aimed at enforcing strict laws or a code of behaviour, as well as fostering sustainable production by promoting car recycling and catalytic converters.
Investment strategies, research and development projects of companies were affected.
In 1993, FIAT carried out the F.A.RE Project (Fiat Auto Recycling), promoting a car recycling project or "cascade recycling" of some components such as propylene bumpers, that become raw materials through crushing and granulation processes and were used to manufacture simpler components (air channelling devices) in the first generation, or car mat in the second generation. Recycled glass was used by other companies to manufacture bottles and jars; seats were used in the building and decoration industry to make fitted carpets. In the third generation, or the last level of the "cascade", materials already exploited and not to be recycled again were used as alternative fuel for furnaces, recovering their residual fuel potential.

nanodesign > Nanotechnologies open up research into complex objects invisible to the human eye, such as micro-robots able to repair tissue damage, to destroy pathogenic micro-organisms or to impede or even invert the mechanisms of ageing. While in biology nanotechnology is associated with the genetic engineering revolution, in the science of materials the prospects are equally fascinating. Chemistry, physics and electronic engineering all find their maximum interfunctionality in the possibility of generating materials and devices whose electrical, optical, thermal and mechanical properties can be pre-set. For example, the application of special devices, such as organic Leds, made up of an organic film placed between two conductive layers and able to convert an electrical signal into a luminous one, could offer greater design freedom in the lighting sector. Other examples include organic polymers suspended in vitreous or plastic matrices with special optical properties for the creation of intelligent glazed surfaces to be used — once costs have been reduced — in the automobile industry, building construction and agriculture. Nanotechnology can, therefore, be considered a form of matter digitalisation ushering the science of materials into an evolution analogous to that triggered by the digitalisation of information.
The axiom Less is More assumes new meanings as compared with its origins. It becomes a search for a new balance between artifice and nature through a reduced use of materials and energy and a greater attention to the creation of eco-compatible objects; it becomes the skilled control of the most micro-metric tolerances in production and the stimulus to reduce redundant and invasive formal notions; it becomes the stimulus to design and manufacture sophisticated technological objects using tiny and immaterial information technologies, intelligent in their functioning and expedient in their interaction with the end-user, whose access involves simple, ergonomic gestures and no expert knowledge; finally, it becomes an opportunity for the study of the reduction of the physicality of things within the scenario of the application possibilities of nanotechnology — i.e. the extreme reduction of form.

[2]
p.97

Innovation through culture

social chromatism
from Il colore racconta tutto il design, DIID 53|2012

semantic value
from Design fra identità e diversità, DIID 42–43|2009

design schools
from Design for Health, DIID 27|2007

conjugation
from Lazio Regione di Roma: sistema design, DIID 21–22|2006

productive territories
from Consiglio Italiano del design, DIID 26|2006

italian style
from Comunicare il design italiano, DIID 49| 2011

italian look
from Forme del Made in Italy, DIID 10–11|2004

Italian design is an intricate heritage entirely consistent with the complex identity of its people and the diversity of its territory. Italian design, however, is a broad and solid social phenomenon, in which non-transitory values connected with the past converge, by way of a common intellectual programme that may be summarized in the expression "innovation through culture." It is a phrase that can be translated into a programme, the one most adhering to an idea of continuity between different generations, that makes the work of masters, artists, young designers, anonymous designers, and businesses all connected in demonstrating the original material culture that history has handed down to us.

social chromatism > From the mid-1990s onwards, it has introduced a new, indispensible relationship between technology and colour, initially with the use of transparent plastic in various hues, and later with a focus on a single, bright shade of translucent white. It is no surprise that Apple's most recent patents cover not only technology but also colour.

First and foremost, the origins for this affirmation lie in the wellknown observation that from the 1960s, design in Italy — as an independent, self-referential discipline — was a result of crossovers in the arts, the osmosis of design experiences and the overlapping of expressive languages. It was a hub of styles, forms and types, of industry and craft, and above all of applied and conceptual art, which has never been afraid to embrace colour. Indeed, with the later *radical* movements in the 1970s, colour became a sort of social and cultural manifesto to depict the experiments underway with industrial products, with an array of artistic languages introduced by postmodernism, which was reflected in the pop culture of mass communication and consumption.

The projects of Alchimia and Memphis may be offered up as examples all too frequently, but they remain unparalleled. They were similar and different at the same time, whether the item in question was artificial or natural, a graphic colour or a coloured fibre. With their work, colour became a performing, narrative material.

Colour became a way of defining what was essentially a new discipline of design with Andrea Branzi, Massimo Morozzi, Clino Trini Castelli and the *Colordinamo* research between 1975 and 1977. Stubborn colour projects established the values of "design primario" ("leading design"), which included the "soft qualities" of products. As eclectic and ambiguous as it may be, there is a "pictorial" dimension to Italian design. It is dominant in some cases, especially due to the metaphysical changeability that can be attributed to industrial products. Furthermore, the image of "painting" introduces the key idea of reducing objects to two dimensions. It is only a conceptual metamorphosis, but it is enough for the valentine "red" written on sheets to be portrayed as a poppy red pantone hue.

semantic value > The identity of Design of the Italian school is unique enough to be studiedin phenomenological terms. Industrialisation in this country wasn't affirmed until the 1950s. Yet, already in those years, Italy's Design was becoming known at the global level. An extraordinary short-circuit was produced between the so-called economic boom and the context of a widespread and unique architectural and artistic heritage.

On the one hand, there is a system common in areas containing small and very-small family-owned companies, where we find a multitude of complementary artisanal skills: this is what economists studying the "Made in Italy" phenomenon call the "system of industrial districts".

On the other hand, there's a widespread culture of Design formed in technical schools and academies, in Architecture Schools and in Polytechnics, but above all, through daily contact with the "beauty" that is so abundant all around us, and which creates "good taste" to such an extent that it has become a part of Italians' genetic code.

Creativity is associated with deeply-rooted good practices of the trades that are passed down from generation to generation. These trades were born around the big Roman factories and then the Medieval and Romanesque ones, then the Renaissance ones of Brunelleschi and Bramante and the Baroque ones of Bernini and Borromini: stucco workers, modellers, turners, tailors, upholsterers, marble-workers, fresco painters, stonecutters, sculptors, painters, decorators, tinsmiths, carpenters and unparalleled artists, all serving emperors, kings, princes, dukes and marquis, plus popes and cardinals.

The great patrons at whose courts were produced imposing and magnificent works of art, but above all where the practice of creating excellence using excellent artistic and technical skills was fed. The new trade born with the arrival of industrialization processes in Italy, that of the Designer, is measured against this background, with aheritage that is part of the country's material culture and in the boom years arrogantly enters production places from which objects of extraordinary fascination exit; objects that are desirable for their semantic value even before their functions, for the emotion they transmit, for that familiarity that they immediately evoke, colorful and soft in their smooth lines, Euclidean in their lines and in their geometries, made of complex manufacturing systems or of simple assemblies, they are desirable objects to such a point that the new Fiat 500 was considered the world's sexiest car.

In Italian Design, imagination is paired with rigorous designs that mix functionality with irony, and emotion with the complexity of the balance between the aesthetics of the artefact and careful attention to research into new materials, appropriate technologies and all that goes into the most innovative experimentation, with a view to the future. There are many great masters of the Italian School and each one is unique. I think that here it is appropriate to remember Bruno Munari. A man of rich and extraordinary complexity, whose experience crosses Visual Arts and Design with the spirit of the *homo faber*, but with the culture of an intellectual who had frequented applied arts and sciences, philosophy and pedagogy, literature and great travels to the Orient. He always focused on applying method to his creativity, which allowed him to create proposals for men and the world of children. In the Italian Design school—also supported today by a network of institutions inside and outside the national university system, at the leading edge of European education—there is a methodological and cultural thread that helps us adapt: major economic transformations or major cultural movements that follow each other propose new artefacts profoundly rooted in this country's history.

design schools > The increase in graduate courses and faculties of design is a response to the demands of young people who want to become designers, but above all, it caters to the needs of companies who require new kinds of professionals not trained by traditional faculties. In fact, design schools are the ones (perhaps the only ones) which, compared to other schools, have modified the way they teach to adapt to the changes introduced by new technologies.

The IT revolution has changed the way we access knowledge; it has transformed design techniques, making it possible to produce perfect simulations; it has changed the relationship between design and production through remote control of numerically controlled machines which implement with pinpoint precision the indications contained in a design; it has introduced automated production processes instead of traditional manufacturing machines, it has modified the relationship between the object and the user by introducing products that modify their performance and even their morphology through an interactive relationship with the user.

All this is taught in design schools: not just the theory, but the practical issues as well; the boundaries between theory and practice are established by these new fields of learning. In fact, in design schools, teaching now exploits a multidisciplinary. In short, design schools have quickly understood that they need to teach students new professions given the enormous influx of artificial material and immaterial objects which require new knowledge, new understanding and new technical and operational skills.

In design schools, the entire training course revolves around a "core issue" which is based, almost deontologically, on the potential of the innovative contribution of computers, on the way they have revolutionised the way we think, express ourselves and communicate, on the way they have radically changed production processes and the way we interact, use and consume products.

conjugation > There's no doubt that the production system of Lazio, the Region of Rome, certainly plays an important part in Made in Italy.

First of all, it's important because it combines tradition with creativity and innovation, in other words, the three crucial ingredients needed to create excellence. It's important because of the extensive growth of industries that have turned the opportunities and traits of a region into a system. The industries have merged art and technological innovation; they have taken their historical heritage and used it to generate the potential expressed by the new professions that have developed in the field of tourism and culture. It's important thanks to well-known brands or designer products, like the collections of famous designers: Valentino, Capucci, the Fontana Sisters, Biagiotti and the up-and-coming designers like Galante, Grimaldi and Giardina.

It's important because of the anonymous products that satisfy our everyday needs: the products made in the industrial district of Civita Castellana, or in the composite industrial system complex that has sprung up between Pomezia and Latina.

It's important because of the hi-tech artefacts that stand up to global competition, for example, the products developed in the telecommunications and aerospace district. It's important because of the potential of its chemical, pharmaceutical and biotechnology industries.

In the world of small and medium-sized enterprises, there is a cultural system that is a source of added value in the productive reality of the Lazio Region. Furthermore, the educational and training system as well as the network of technological poles and centres of research and experimentation supports the productive system by introducing innovative processes and products.

Finally, it's important because the region has its own specific traits. Apart from its manufacturing industries, it has a widespread network of enterprises that operate in the service sector or produce immaterial goods that exploit the skills of new professionals, for instance, the Audiovisual and ICT districts.

Promoting culture by enhancing cultural and archaeological assets in the entire Region is also an economic undertaking: for instance the activities of the prestigious City of Music in the Rome Auditorium and the Academy of Santa Cecilia or those of the widespread and complex network of museums in Rome and the region. SMEs have developed around these activities. These companies employ professionals from the world of design, communications, graphic design, publishing, multimedia, exhibit design, public design and special effects design. In some cases, professional figures have developed in associated businesses that have grown around the Italian State television, RAI, and/or consolidated traditional realities like Cinecittà or the Opera. The culture and production of stage-sets, costumes and clothing design are internationally renowned and there are Oscars to prove it. The roots of these professions are stored in the Warehouse of the Opera House which, apart from the stage-sets, houses over 50,000 costumes: they bear witness to Italy's artistic history. In other cases, new activities and enterprises have developed in the design and production of merchandising, or structures like Enzimi, Zètema and Civita that promote and organise events to enhance our cultural heritage or hold art exhibitions alongside the Macro and Maxxi museums that are part of a museum system that has become global.

In the field of design, to satisfy the requirements of the many, sometimes new professional profiles, the La Sapienza University in Rome offers a whole range of educational courses: three-year graduate courses, specialist courses, masters and doctorates.

Apart from the universities in Lazio involved in the field of design, there are other public and private schools which, in some cases, have very high standards (IED, ISIA, Istituto Quasar, the Academy of Fashion and Costumes, the Academy of Fine Arts, Koefia, etc.). The educational syllabuses of design courses reflect the wide range of required subjects which depend on the increasingly complex nature of production. Industrial designers, in fact, have to be versatile: they're not just people who look after the aesthetics of a product, they are highly qualified specialists. Courses include the design of material goods, infrastructure, services and immaterial commodities. I'm referring to visual communications, multimedia and interactive design, i.e. a design that uses the right styles and tools and the necessary and feasible technologies when involved in the graphics of editorial products, packaging and coordinated brand products, animated images and iconic interfaces for computer networks.

Furthermore, with the introduction of microelectronics and the widespread use of computing, telematics, robots, nanotechnologies and new materials, industrial products have acquired performance-related and marketing traits that require multiple skills, to such an extent that design and production methodologies are constantly being reviewed and updated. In fact, design has to be planned and administrated vis-à-vis production strategies and marketing requirements: this is the task of design management and design direction. It's clear that the teaching requirements are in line with the work carried out by the network of centres that create a regional research system: a system that includes strategic projects developed by Space Agencies, the work of the Galileo Test Range (GTR), the work on new materials, nanotechnologies and their function and the research on the implementation of laser technologies in advanced diagnostic equipment. In the Lazio region, there is a very high standard of design culture and project culture used in the production of material and immaterial artefacts: the added value comes from the relationship that creativity and innovation establish with tradition.

productive territories > Italian design is no longer unitary and univocal, but extremely versatile and multiform; it represents different specialities and vocations from all over the country.
Italian design is now an integral part of a wide range of activities and is used in many fields of production and industry.
One successful example of design are the aesthetic and technological products produced by that amazing experimental workshop, the Ferrari factory in Maranello, as well as the entire Motor District in Bologna.
Design is used in Trieste, in the shipyards of the Fincantieri Company to build the huge transatlantic liners commissioned by shipowners from all over the world, or in other shipyards (Gruppo Ferretti in Ancona, Forlì, Torre Annunziata, Fano, Cattolica, La Spezia and Sarnico; Gruppo Rizzardi in Fiumicino, Sabaudia, Gaeta and Posillipo, Gruppo Aicon in Messina).
Design is used in the Research Centre of the Indesit Group (household appliances)-part of the history of Italian design-and the entire Engineering District of Fabriano-Jesi where there are still some outstanding companies, like the Elica Group.
Design is used in the manufacturing district in the north-east, in the fields of engineering and subcontracting in which Italian SMEs are world leaders in the production of manufacturing equipment for the textile sector or precision instruments.
Design is used in the state-of-the-art labs that research nanotechnology in order to apply innovative solutions to extremely small components, for instance in the Veneto Region (Veneto Nanotech), in Lecce (Apulia Region) and Modena (Emilia Romagna).
Design is used by ASI and Alenia; this has led to the development of the Aeronautics and Aerospace districts in the Lazio Region, as well as in the interregional districts of Piedmont and Campania.
Design is used in the audiovisual sector in Rome, the first step towards increasingly advanced research.
Design is used in exhibitions, stage-sets and costumes, carrying on a tradition which in Rome has won international acclaim and more than one Oscar for production design.
Design is part of the fashion world in Rome, Florence and Milan as well as the rest of Italy. Apart from the great historical couturiers, Valentino, Capucci, Armani, Versace, Ferragamo, Biagiotti, Prada, Missoni, etc., there is a network of wellestablished and productive ateliers all over the

country. It is one of the most important and successful production and product systems in the world.
Design is used in all the more typical fields of Made in Italy.

italian style > Communicating Italian Design is an arduous task, even more so when you have to do it in a few magazine pages. But taking stock, even if not in an exhaustive way, every now and again is useful and appropriate.
To do this we have chosen a specific period of reference: the first decade of the third millennium.
This is a decade in which the Italian design system became increasingly polycentric; an expression of a range of productive and cultural realities which have affirmed their specific nature and special features (although in an increasingly globalised manner), for example in the area of training, with various schools forming an extensive network across the country.
It is a decade in which new designers emerged, linked to the Italian production system, whether from Italy or from different countries.
In some cases their way of designing carried on the theoretical, cultural and methodological approach of their teachers, who contributed to making Italian Style a worldwide success. More often, there was a trend of discontinuity, and a new way of understanding the discipline of design.
A new way that accounts for the elimination of distances between national cultures and traditions which were still far removed in the previous century. It is a new way of looking at the present and the future, with awareness that the boundaries of art, design and architecture are becoming increasingly ephemeral, as there is growing hybridisation and contamination between these disciplines and new material and immaterial expressive forms of communication are emerging.
It is a decade in which, as well as the traditional areas of design (furniture design, interior design, product and transportation design) other applications emerged like exhibit and public design, fashion design, visual communication and multimedia design, etc. These were ten years in which the interest of the media in design grew-suffice it to think of the columns dedicated to the subject in many magazine supplements of daily newspapers.
It is a decade in which many institutions were set up with the aim of spreading design culture: from the Musei d'impresa in the Farnesina Collection, to the Fondazione Valore Italia at Milan International Furniture Fair, from Abitare il tempo to the opening of the Italian Design Museum at the Milan Triennale which after having answered the question "What is Italian design?" in its three exhibitions, Le sette ossessioni del Design italiano, Serie fuori serie and Quali cose siamo, has provided a new answer only recently in Le fabbriche dei sogni, the fourth exhibition that presents uomini, idee, imprese e paradossi delle fabbriche del design italiano (Men, Ideas, Businesses and Paradoxes in the Italian Design Factory).
Despite a great deal of attention focusing on the more traditional applications of design, too little attention has been paid to design that is expressed in high performance products; those in which innovation (and not only formal innovation) involves innovation of type, of new products for new

needs or of so-called anonymous design.

Despite plenty of (well-deserved) attention focussed on leading figures in the design world, too little space is given to emerging designers and new design, involving different areas of the home, the interstitial spaces between man and his environment and man's needs and the emotional involvement of the senses; or involving the relationship between man and the object he uses, as part of new aesthetic values and new visions of the world he desires.

We still remember and remind ourselves that the age of Italian design is the result of history-not that kind of slightly pompous and self-referential history, which is often represented, but that which Giulio Carlo Argan recounted in his extraordinary 1982 essay "The Design of Italians" written as an introduction to the exhibition "Italian Re-Evolution. Design in Italian Society in the Eighties", where he clearly explains the special features of Italian Style in design and where its roots are to be found. This is the reason why it is once more presented in this issue of DIID.

We have described Italian design in this first decade by focussing our attention on two of its special features: the genius of its know-how and the iconic value of its products. We have recounted how the creativity of 'Italian Style' is translated into product proposals, as a result of genius and know-how and how these two features combine the culture of artisanship and tradition with that of innovation in the design of products, or how iconic value — a special feature of the objects that make up the history of Italian design — is still an added value in the products of the last decade.

italian look > In 1982 the title of the introduction written by Giulio Carlo Argan for the exhibition catalogue "Italian Re-Evolution. Design in Italian Society in the Eighties" was The Design of the Italians. A title that implicitly defines the subject and the context of his critique.

This seems like a good starting point to try and understand the Made in Italy phenomenon, because it's based on the critical awareness that Made in Italy has its own cultural specifics and is not an overrated instrument of cultural promotion that now needs a comprehensive re-analysis. That's why we think that it's useful to study the Made-in-Italy phenomenon by focusing on all the technical, aesthetic and economic aspects of an industrial product: the same topics that Giulio Carlo Argan examined in his book.

At the end of the second world war Italy came out of its cultural isolation and re-launched an industrial production that didn't focuses on an abstract society, but on real society, avoiding any apologia to ideology. The small, budding enterprises all had one thing in common: the ruthless typological re-invention of everyday objects and, as Argan says, the search for their iconic visibility and the ability to stimulate the senses, the sense of touch and sight, in short, man's senses.

This is the approach that the Italian look adopted and fostered. In essence. its growth depended on giving these products visibility. Even when sophisticated technology made them high-performing tools, these products still retained their formal characteristics: everything tended to be visible. Even rooms such as the kitchen and bathroom — traditionally neglected by aesthetic research and based on efficient technical systems and functional rationality — began to betray signs of product quality that varied and personalised these spaces as well as making them socially acceptable.

Gradually new materials began to transform rooms, objects and their shapes. Flexible, extremely lightweight plastics, transparencies and vivid colours changed our relationship with objects and our tactile and visual sensations compared to what we were accustomed to. Objects began to loose their material value. On the contrary, their ephemeral dimension represented a new added value.

The specific nature of Italian design began to falter with the delocalisation of style and the hybridisation of cultures into the global style of remote objects. Above all, it faltered when the function of design changed, when design was used not only to give an object its shape, size and weight, but, by using computer technology and communication systems, to change the way and speed with which people socially interact. This crisis matches the decline of Italy as a country, its capacity to progress in the fields of culture and science, art and technology, economy and politics. An important crisis of industrial production, despite a common, shared feeling that refuses to entertain the idea of decline by putting on celebrations, protectionist measures and talks about re-launching Made in Italy. In other words, something that is haphazardly associated with agricultural or food products that represent the old traditions of Italy and its culture These are the synthesis of the amazing professional and disciplinary abilities that come from different cultures and countries and which in Italy are only produced or assembled. So I ask: is Made in Italy something that this country can do by using our traditional talent, culture, creativity, initiative and resources? Or is it just a brand that gives products an added value? Should we root our design in local cultures and traditions or insert it in a complex network where innovation and information have no limits? Is the concept of Made in Italy compatible with the global dimension of the free circulation of ideas and the so-called delocalisation of production?

§
[2]
p.106

Traditions and transformations

a lesson
from Il design è ubiquo, Lectures 1, Roma 2014

units for architecture
from L'intelligenza desiderabile, DIID 58|2014

new ferments
from Design fra identità e diversità, DIID 42–43|2009

between two lands
from Le Pasabahçe, DIID 40|2009

popular design
from Design for Health, DIID 27|2007

aestheticising
from L'ornamento e l'estetica degli oggetti, DIID 23|2006

no-name
from Mass design o il potere dell'oggetto anonimo, DIID 15|2005

the flow of products
from Contaminazione, trasferimento, DIID 14|2005

cultural complexity
from Identità e diversità, Mercosur Design, DIID 12|2005

Recollections and testimony, transmitted between people, between different generations, as well as social events, rituals, shared practices handed down. All this is tradition. Tradition, then, while consolidating itself through cultural sedimentation, changes without interruption, because it transforms its own knowledge. It updates it continuously and randomly. It is the same process that accompanies the exegesis of design: being a place of deeply rooted memory, and at the same time placing oneself in the groove of change, of alteration, of modification-spontaneous or induced by causes inside and outside its own mandate. Design is therefore a model of equilibrium between the profundity of the past, the condition of the present, and the aspiration for a possible and useful future.

a lesson > Always looking for a good read to generously facilitate the task of writing, we ran into a 1917 essay by Russian philosopher and mathemetician Pavel Alexandrovich Florensky, with the title "Lesson and *lectio*." He says:"It would not be out of place to define the ideal lesson as a sort of dialogue, of conversation between very spiritually close people. A lesson is not riding a tram that carries you, inexorably, on fixed tracks and takes you to your destination by the shortest route, but a walk, an outing, albeit with a precise, final point, or better, a walk that has a precise general direction without having the single declared need of arriving there or of doing so via a specific road. For those who walk, it is the walking and not just the arriving that is important; those who walk proceed serenely, without quickening their step. If they are interested in a rock, a tree, or a butterfly, they stop to have a closer, more careful look. At times, they look back to admire the scenery or — and this also happens! — they retrace their steps, remembering they have neglected to fully observe something instructive."
Although redundant and weighted with emphasis, this is a passage that establishes, with agreeable moral acceptance, the basic principles of teaching. The lesson, is the dialogue between "very spiritually close people".

units for architecture > In recent years, the field of architectural design has adopted the word "smart". It seems to have become a remedial term used to assert one's support for a sustainable and integrated modernity that makes cities more usable and connected, replete with myriad online services. Nevertheless, all this is not enough. The impressions that the "smart" concept evokes often prove to be rhetorical overstatement, especially given that when we consider the concrete examples produced by smart culture — both in terms of the buildings and the products that use new communication technologies no suitable models for new forms of artificial intelligence emerge, forms that could allow us to glimpse the physical structure of new aesthetics for different needs. In short, scientific knowledge moulded by architectural design seems more of a slogan than an actual opportunity for reviewing established design models, or testing new product and service paradigms, or investigating the reasons behind a design as it develops into a product. I visited the International Architecture Biennale of Venice, edit by Rem Koolhaas, dedicated to the theme of the "Elements" of architecture. After several editions dedicated to the celebration of "contemporary", this exhibition was intended to investigate the foundations of architecture, starting from the unities of "model and type" (a staircase, a window, a ceiling), which have always been a critical and physical repertoire for designers.
Observing what has been exhibited, I got the feeling that the transformation of materials and technologies show those changes of the expected paradigms "to think at the new with the old".
If we take in consideration the classical "case" of the door, we would know that its functions, in terms of absolute security, are now being invaded by invisible systems, such as metal detectors, which have rendered useless (certainly in specific contextual conditions) its physical status.
And then, about the theme of the "floor", it is no longer just a footpath but an active configuration

ables to produce heat or cool — as demonstrated by the Rotterdam's "techno-disco" that produces energy — or as a system that guides the Robots thanks to magnets, as happens in some storage areas of Amazon where the human presence is forbidden.
For example, in the area dedicated to the "fireplace", it was possible to observe that with digital technology "things" become more responsive to human needs.
On one side, a the primitive fireplace was showed-one of the first human solution to control the fire discovered near Valencia, Spain; on the other side, there was the future of home heating, a project edited by the Senseable City Laboratory MIT Lab. Thanks to the tracking technology, we not need to heat the whole room to avoid the cold, but instead, we can have a "simple" heat bubble that constantly follows and wraps us in our every move.
According to the opinion of the US researchers, this result is based on an ideological approach to "computer science", the only one that allows us to freely speculate on space and services.
The exhibition *Elements* does not stigmatize the inviolability of the past, neither uncritically supports the present, but simply offers indications for future developments.
Obviously, among these indications, the principal is about the infiltration of the "digital" (as synonymous of smart), into objects or systems, big or small, that are allowed to "hear", "talk" and "transmit".
Nonetheless, understanding and "admiring" this process of transfiguration of any things into animated systems, doesn't open the question about their usefulness, but rather about their desirability.

new ferments > The whole scope of social involvement remains of the MM experiment, but the approach is different for the relationship between Design and the nature and specificity of the contexts which become not only references for the use of the materials, but above all the fundamental inspiration for the figurative and formal solutions of objects; in nature shapes are sought to inspire new aesthetic Designs.
The Scandinavian school may be the one whose identifying traits still have the strongest link with its origins and its founding fathers (Alvar Aalto or Arne Jacobsen) for whom it was almost a deontological commitment to both affirm the relationship with one's own tradition and to affirm a functional and organic conception for which the object is conceived with respect to an architecture and a dialectic balance between humans and the environment, and humans even become the creators of its assembly and of its composition in the domestic space.
Because of their different histories and culture, and because of the different ways and times in which the countries were industrialized, the Mediterranean school has a more diverse identity system which is different in Portugal, Spain, France, Italy and Greece.
The French school is doubtless the most deeply rooted in the great movements which were the origin of *Industrial Design* at the turn of the 20th century.
What's more, Paris is the point of reference for the greatest expressions of the artistic avant-garde

and the economic system measured itself with the industrial revolution very soon in France.
A School of Design deeply rooted in MM's culture and theories of rationalism and functionalism was established. Design is expressed through great interest in society and people, and in technological innovations.
This is seen in the importance given to decorative arts shows for thepresentation of new products, and specific schools, such as *Conservatoire National des Arts et Métiers* where Jean Prouvé taught.
Design is not only a factor of technological development, but also a fly-wheel for the productive system, and an opportunity for contemporary people to live in a new and better society, as expressed by Le Corbusier in *Esprit nouveau* or in his concept of the home as a *Machine à habiter*, a bona fide manifesto of modernity.
The French school's identity is built on this foundation, on the experience of many masters who, like Jean Prouvé, in the last century, expressed the ability of combining craftsmanship and entrepreneurship with the spirit of the *homo faber*, ever inspired to construct objects through the reasoned use of materials, with a style that combines them to exalt the functions of the objects but to give them a formal expression, to translate technicality in aesthetics. These and others are the major works — by Rene Herbst, Robert Mallet-Stevens, Eileen Gray, Sonia Delaunay, Charlotte Perriand, or Pierre Chareau's *Maison de Verre* — which constitute the humus that gave the French School its identity. The *Maison de Verre*, Pierre Chareau's only work is also the representation of how the intimate relationship between industrial objects, figurative works of art, and living spaces with no other references if not the idea of modernity, can be combined in a sublime representation of industrial civilization, in an extraordinary expression of modernity.
Since interest in Design in Spain and Portugal exploded only at the turn of this century, the Iberian School still needs to consolidate into a single system the many ferments characterizing it.
But we can say that in this region Design has a direct relationship with the Post-industrial cultureand economy, almost skipping the industrial ones, or at least experiencing them but not making them its own.
The new ferments are expressed not only in products of great interest but also through a widespread sensitivity to the influence of local cultures, production and consumption, against the background of global impact; by a sensitivity to a concept of Design not only as a discipline linked to objects and their function, but as an instrument to explore processes, often to serve science; by a sensitivity to the need to produce a new way to spread new things, an ethical Design in today's change of epoch and methodologies, representing the guiding star of social and sustainable development of cities and consumption.
But it is also in its roots of the great handicrafts tradition and in the use of applied arts that, in the past, the very image of the urban contexts and domestic environments was successfully characterized with strong expressive content through the use of high-quality products such as the multicoloured *azulejos*, the ceramic tiles covering buildings, or in the two-toned *calçada portuguesa*, the traditional white and black basalt or granite pavements in Portuguese cities. These show the

quality of the Portuguese urban space. And, further, the roots in tradition of the modernist culture have given Spain works of great importance. Yet, it is clear that a highly evocative Design is making inroads especially in Spain: products that stimulate all the senses, products that capture life but that draw the hand. Joyful objects that create emotion are the result of new ideas, of innovative and avant-garde suggestions, that express all the radiance of the Mediterranean.

between two lands > A different Mediterranean and different Design. We want to investigate the Design experiences which, as an expression of material culture, are to be found in the part of the Mediterranean that is considered peripheral by the Eurocentric view of history and culture and that I, on the contrary, believe is central, precisely due to its history and culture. First and foremost, the centrality is geographical, because Istanbul is the only city in the world that spans two continents: Europe and Asia. It is also based on Istanbul's political, economic and social history, first as the capital of the Byzantine Empire and then of the Ottoman one. Istanbul is central as an extraordinary place that conveyed its unparalleled cultural experience to the entire Ottoman empire; a place whose true greatness is not always recognized. We spend a great deal of time studying Lorenzo the Magnificent and the Italian Renaissance, and less looking into the exceptional cultural and political experience of Suleiman the Magnificent and Sinan, the architect he entrusted with designing the mosques that give an unforgettable character to Istanbul and its cityscape, standing out in the overall view of the buildings on either side of the Bosporus and the Golden Horn. Sinan is one of the most important architects of all time. We really should include him in the programmes in our architecture schools and study his work more than we do now. He was a contemporary of Michelangelo and Palladio and for half a century he was behind all of the most important buildings in the Ottoman Empire, either as the designer or the construction supervisor. Edmondo De Amicis and — over a century later — Orhan Pamuk have expressed their thoughts *on* and *from* the Bosporus. The ideas of one are reflected in those of the other, with both discussing the multicultural, multiethnic dimension of this exceptional place. De Amicis does it in a piece of correspondence from the Galata Bridge, Pamuk in a *rêverie* that describes his thoughts and fantasies as a child. One observes people, the other observes things. Edmondo De Amicis tells us how, observing the flow of people on the Galata Bridge going back and forth from one side of the Bosporus to the other, it was possible to pick up whether they came from the countries of the West or the East. Even though they were all part of the Ottoman Empire, there were different, distinguishing physical characteristics, ways of communicating either verbally or by their habits, and features in all of the baggage with them. He saw it as a multicoloured, multiethnic caravanserai; an extraordinary combination of peoples, cultures, religions, usages and customs, involving non-stop coming and going.

In a short piece at the end of the catalogue for the "De Byzance à Istanbul_Un port pour deux continents" exhibition, Pamuk tells us how he looked at a *Pasabahçe* — one of the last of the old steamboats that have been crossing the Bosporus for over fifty years — and was reminded of his feelings as a child as he gradually came closer to the coast, challenging himself to pick out a familiar sight, a distinctive feature of something with which he identified among the plethora of buildings on the chaotic front of the city on the Bosporus, full of roads, large advertising boards and millions of windows. This description of the child's eagerness to recognize his own house or a reassuring identifying feature is a metaphor for the diversity in uniqueness, expressed in poetic form. This is a characteristic of the Mediterranean itself, whose very name means 'in between two lands'. This new definition appeared for the first time in the 7th century AD, following the Islamic conquest of the entire African coast and part of the Iberian peninsula, causing the Roman name *Mare Nostrum* to fall by the wayside for good.

The Mediterranean is just how Fernand Braudel described it:" ... A thousand things at once ... It is not a civilization but a series of civilizations stacked on top of each other. Travelling in the Mediterranean means coming across the Roman world in Lebanon, prehistory in Sardinia, Greek cities in Sicily, the Arab presence in Spain and Turkish Islam in Yugoslavia. It means plummeting through the abyss of the centuries, as far as the megalithic constructions of Malta or the pyramids of Egypt ... the Mediterranean is an extremely ancient crossroads. For millennia everything has been converging there, complicating and enriching its history: beasts of burden, carriages, goods, ships, ideas, religions, lifestyles. And plants. You think that they are Mediterranean and yet — with the exception of olive trees, grapevines and wheat, which are native and established themselves extremely early on — almost all of them originated far from the sea ... Those golden fruits among the dark green leaves of certain trees — oranges, lemons, mandarins ... — come from the Far East and were introduced by the Arabs ... agave, aloe, and prickly pears ... come from America ... the big trees ... which even have a Greek name, eucalyptus ... come from Australia ... and cypress trees are Persian. Nonetheless, these elements have become constituent parts of the Mediterranean landscape ... in its physical landscape, as in its human one, the Mediterranean is a crossroads. In our memories, the irregular Mediterranean becomes a coherent image, a system in which everything blends and reforms in an original unit."

We are trying to understand the Design from the area that looks out on the Mediterranean from the South: another Mediterranean where a large number of differences come together to make a single whole and another Design which expresses a creative dimension with roots in the traditions of a culture dating back thousands of years, in the areas of the Middle East near the sea and the regions of Africa.

Its products, with their mixture of old and new customs, reach as far as the European areas of the Mediterranean. A web with a thousand invisible threads holds together, in a single entity, usages that connect places which are different and distant from each other.

There are many examples. One is anise spirits, which are interesting because the route to their widespread success was through the ports of cities on the Mediterranean. The aroma and flavour are shared by a number of spirits with different names and characteristics: *Tutone* in Sicily, *Pastis* in France, *Raki* in Turkey, *Ouzo* in Greece, *Arak* in the Middle East and *Sambuca* in Italy. It is no

coincidence that the name of this last spirit comes from the Arabic term *Zammut*, as they used to call an anise-based drink that would "dock" in the port of Civitavecchia on ships from the East.
Another example is provided by the spread and development of ceramics production in the Mediterranean basin, starting with the ancient *Iznik* ware that decorated the great architectural works of the Byzantine Empire and then the Ottoman one, with manufacturing techniques, forms and decorative colours that have descendents ranging from *Vietri* ceramics to Portuguese *Azulejos*.
The culture of this alternative Mediterranean has also been a source of inspiration for the contemporary European art scene, which revives figurative materials from the culture of the other Mediterranean, letting itself be influenced by symbolic and expressive worlds, in order to portray new forms and figurative scenarios. Paul Klee, a leading contemporary painter and one of the outstanding teachers at the Bauhaus, is a fine case in point. His multiple visits to Egypt and Tunisia influenced the colours and their equilibrium in his painting, as well as the composition of the shapes; the visual experience of those lands is clear and significant, even though it is altered by abstract methods and processes.
There is a Design scene in the countries on the southern and eastern coasts of the Mediterranean that expresses itself in the engagement between tradition and aspirations for innovation. Rather than appearing in the use of complex high technology, product performance or manufacturing systems, this innovation involves reworking forms and using new materials, especially regarding textiles and their use. Indeed, there is no shortage of successful fashion designers, such as the Turkish Cypriot Hussein Chalayan, who carries out extraordinary experiments and innovation. Other examples are Soumiya Jalal, a textiles designer who lives and works in Morocco, and further to the East the Turkish designer Aziz Sariyer, also known as the founder of Derin. In addition, there are companies like VitrA, whose business ventures have taken it from a nerve centre of the Mediterranean all over the globe. However, there is also innovation that represents the desire to contemplate the future with new ideas, providing a form of expression for the various lands that look out onto the alternative Mediterranean, where local design goes head to head with globalization, albeit with some contradictions and many differences between the countries.

popular design > Spreading the word about design culture not only improves product quality, but also our own quality of life.
Several European countries, for instance in Northern Europe, are excellent examples.
In these countries where design is part of their collective cultural heritage, the characteristics of industrial products lead to widespread quality in public design, interior design and in our everyday life.
Design culture is interpreted not only by famous professionals who have created high quality products such as Arne Jacobsen or Alvar Aalto, but also by many designers who satisfy the public's widespread demand for quality design. This design gives an added value to everyday products: in

public urban areas, in homes, offices, schools and caring facilities. It's no accident that the popular design produced by IKEA was created in the north of Europe.

aestheticising > In the past, ornaments have been considered either as organically linked to the technical and functional traits of an object which gave it a certain aesthetic quality—albeit with its own artistic specificity — or as something added to its shape and function, like a crust. In his book, "Ornament and Crime", written in 1908, Adolf Loos maintained that cultural and social evolution required ornament to be excluded from functional objects and architecture.

The completely divergent theories by Ruskin, Riegl or Focillon are just as famous: Ruskin tended to give decoration an aesthetic dimension and believed that function and technical conformation merely played a utilitarian role. Riegl and Focillon, on the other hand, considered that decoration either coincided with form and artistic workmanship (insofar as they represent man's aspiration to shape space and everyday objects) or with specific narrative principles.

It's true that man has always expressed his desire to embellish the space he lives in, either on objects or his own body.

Balance and symmetry were the primordial rules that governed body ornamentation either through painting, tattoos or even scars. This is true whether decoration is inspired by a mimetic relationship with visible natural forms or whether it is abstractly expressed through geometric patterns.

"Beautification" of the human body is one of man's oldest customs: he either decorated his own body or wore jewellery. In both cases, ornaments were intended to aesthetically set off or enhance a person's natural features and emphasise differences in customs, sex, ethnicity, social hierarchy and age. Beautification is obvious and deliberate; painting and ornament by incision are often executed jointly on the same body. They are used to represent the concept of ornament are a (permanent) part of the human body, as well as the concept of ornament as an ephemeral accessory yet with its own built-in aesthetics. In both cases, it figuratively represents initiation rites, sexual, social and magical symbols.

Other decorative traits involve the deformation or mutilation of the human body, for instance, ornaments in earlobes, labial discs or the cutting or filing of teeth. In short, the artificial transformation of the human body in order to portray a person by changing his natural appearance. Of course, beautification is not only inspired by aesthetics: symbols can represent a person's social hierarchy or his tribal affiliation, a time of life such as puberty before marriage or adulthood, proudly represented by scars. Colour symbolism is also important in pictorial decoration; this invests the ornament with a series of elaborate artistic traditions which may include: intricate geometric patterns (inspired by decorative mythological styles); red and white ochre zigzag patterns; marks arithmetically arranged in order to affirm magical/apotropaici symbolic values; zoomorphic patterns with broken lines; diagonal hachure interspersed with herringbone patterns.

Different methods or techniques are employed to embellish the human body without using ornaments but with prostheses. The latter are used mainly to reshape the bodily symbols of sexuality

based on aesthetic principles that tend to homologate beauty and decoration and are inspired by the myth of eternal youth.
Lately these techniques are becoming increasingly innovative in order to conceal what is done and make it invisible. Depending on the relationship between decoration, the materials and technique, when ornaments are used by artisans, their plastic effects and stylistic intensity can either create close links with its structural and functional parts or a superimposition, a plasticity that is secondary compared to the product's formal, technical and structural characteristics.
Instead, when an artefact is mass produced industrially with specific manufacturing methods (casting, pressure die casting, rotational dies, etc.), decoration becomes part of the object's basic shape and coincide with its aesthetics. Not only when the object's surface is devoid of those typical decorative patterns such as modelled pieces, engravings, notches, but also when ornate decorations are present on flat surfaces as well as on concave and convex surfaces or surfaces created by three-dimensional shapes.

no-name > How often have we tarried in front of the windows of big shopping centres and looked at the huge "display" and range of work tools like nails, screws, hooks, hinges, pliers, bottles of glue or electric switches? Or other time-honoured objects whose elementary functions — use, technology, physical and mechanical principles — have not made them obsolete, such as scissors, nail clippers, clothes pegs and hangers, hair pins, staplers, brushes, combs, razors, sponges, soap, etc.? Objects with an anonymous design destined for mass consumption and distributed without the help of advertisements. Objects that invade the scenario of the artificial with ostensible discretion, but which in fact play a major role in polluting the quality around us. We're reassured by the fact we're so familiar with how they work and their often unchanged shape or packaging. Or we're curious about the invaluable improvement in their technical performance that apparently doesn't radically alter their shape, but does affect their technical traits (increased resistance, longer life, greater safety which, from time to time, gives rise to different formal characteristics) or the type of materials used.

the flow of products > A designer has much more freedom when he designs and produces an industrial product than when he designs and produces a building or a city district, i.e. the context in which the object will be used. An industrial product is atopical: it is designed to be mass-produced and isn't linked to any particular place. It is intended to be used by anyone in the world. Instead, an architecture or an urban element are rooted to the site and interpret its nature. However, a network of relationships if established between an object and its context-considered respectively as an industrial artefact and constructed space: it is these relationships that need to be critically assessed. I believe that it is these intricate relationships between space and the flow of products that pass through it that provides us with an interpretative key to our modern world.
When you design an architecture or part of a city, you fulfil a mandate; you are limited by a budget and by the fact that the site has to accommodate traditional buildings such as schools, churches, the

town hall, offices, theatres and houses.

At urban and territorial level, these limits are dictated by the relationship between residential buildings, facilities, industries and infrastructure systems. In short, the budget, the site, the typologies and the models are the building blocks that the designer uses based on his own technical skills: the project renews what already exists and proposes what is plausible.

Vice versa, when you design an industrial product, the functional characteristics of the object are all-important, as are the production methods, the evaluation of the intrinsic possibilities made available by contemporary science and techniques to improve the qualities of the product and earn a bigger market share. The logic of industrial production is to always offer an object before there is a formal demand by the user. It also involves inventing the necessary technologies and production methods: an industrial design project proposes what doesn't exist.

cultural complexity > As democracy is gradually established in the so-called "developing countries", design is experiencing an experimentation that few had foreseen. After all, in recent years the growth rate of the GDP of these countries is, on average, much greater than that of European Member States.

This is quite common in Asia (in the first place, China and India), but it is also fairly widespread in Latin American countries despite the recent economic crisis that has involved some areas of the continent. It would be interesting to analyse this phenomenon from a political, economic, social and cultural point of view, but instead we have decided to focus on how this affects contemporary design in Latin America. By design, we mean the ability to invent and manufacture what is artificial, in other words, all the material and immaterial artefacts that are produced to satisfy both the demands of yesterday and today.

Creating Mercosur was a way to establish a common market in this vast area known as Latin America, but this continent is a mixture of very different cultures, traditions, economies and even languages (Spanish and Portuguese). When we analysed the range of design styles spread across this huge geographical area, we found that diversity and identity were certainly the most interesting issues to explore.

The influence of the European model in the field of design — as in other cultural fields — is very dominant in Latin America. However, interest in autochthonous forms of expression is growing by the minute. In fact, increasingly, experiments and productive activities are picking up on local traditions and the cultural complexity of Latin America, even if globalisation remains an omnipresent presence.

The Ulm teaching method has affected — and in some cases still affects — the way in which designers are trained in schools in Latin America, for instance at the School of Industrial Design (ESDI) founded in 1964 in Rio de Janeiro.

This influence has provided method and discipline but it has also created a design concept based on the ethics and style of modernity; this has influenced training programmes, but hasn't changed the specific demands of the local market or the need for enterprises for create an ad-hoc growth model. It wouldn't be wrong to say that the delay in technological and scientific development in countries in South America hasn't helped.

Perhaps it's this common reference to the European tradition of design that's the glue behind the identity of the whole continent. Its matrix has gradually become "hybrid," perhaps due to the influence of local culture in the world-wide globalisation process as well as the changes in Latin American countries that began in the eighties and have continued ever since. These changes include the revival and reinforcement of the full potential of the continent's traditions and culture that have finally exploded in a burst of creative independence, the true expression of all the peoples of South America.

On the other hand, the multicultural and multiethnic traits of the continent, the seamless borders between the countries — like the market that seems to come together in Mercosur — and the lightening speed with which communications, social phenomena and cultural styles travel from one environment to another, means that diversity and identity tend to gel in "hybrid" forms.

In any case, the connection between the design of new objects and tradition is now permeated by the interesting, elegant method of functional transfer, i.e. an exchange of the typically material characteristics of a product from one type to another. So the traits of a new product for the workplace inspire new health products for domestic consumption; the characteristics of objects traditionally used in the market, commerce or religious ceremonies are transferred to everyday objects used in the public areas of our cities. Rarely does this produce vernacular objects because in general there is a tendency to reinvent and not to repropose. It doesn't just involve reproducing objects, shapes or materials that are typical of local craftsmanship; it's more a refined, cultured and ironic reinterpretation of tradition and history. Transfer is also fully exploited in style and formal research: the colours of a crowd during a procession, in a marketplace, natural materials, traditional costumes, etc., are transferred into the new objects. Geometrical designs and the decorative forms of ancient iconography are used as reference in these new proposals.

Another characteristic element of the design approach in Latin America, or at least in the countries with the highest growth in GDP, is self-production. This is extremely interesting because it's an identity factor that in fact preserves and produces diversity. Its diversity lies in the product, in its link to the history, culture and materials that belong to the various contextual geographical, economic and social conditions; its identity in the method: i.e. self-production. A system rooted in tradition; one that goes further yet never involves the production processes that have historically determined the growth of advanced industrial models based on large-scale production for a single, globalised market. In fact, to produce what he designs, to produce it with the materials, technologies and tools at his disposal, the designer becomes an entrepreneur. However, he does all this for a market that basically limits the number of pieces to be produced and allows the designer to dream up new ideas for new products, in other words to be at the centre of a virtuous circular system that continually solicits him for new ideas. This sets him apart from both traditional craftsmen who use age-old manufacturing techniques and product types as well as from industrial production where parts of the process (design, production, commercialisation, etc.) are very specific and are carried out by different actors. Self-production has teamed up with the use of recycled materials. This is a widespread trend and the Campana brothers are certainly the forerunners of a research that has led the field; so much so, that young groups such as the Notechdesign have adopted it as the manifesto of their design work. It's an interesting research that recuperates the homo faber dimension of designers, their manual skills, inventiveness and creativity. Quite apart from its symbolic meaning vis-à-vis eco-efficiency, it often gives the product an artistic dimension, a knack at involving all five senses.

Self-production is certainly controversial. Many consider it postpones the creation of national industrial systems or their ability to measure up to industrially advanced countries in the field of competition and innovation.

However, the experimentation in design in Latin America is undoubtedly original and innovative; it includes the design of products, fashion, graphics, visual communications and interior design.

§
[3]

Compendium

Introduction

Over these years, I have brought life to several publishing efforts, such as *Quaderni del Dipartimento di Pianificazione, Design, Tecnologia* dell'architettura, or the collection of Lectures by instructors in the Department I led.
The group of professors with whom debate was most intense took part in this activity with essays that contributed towards interpreting the present with regard to designing-related issues in Architecture and Design. The set of selected texts that I drew on as I reread a certain number of their contributions became a sort of narrative counterpoint, a compendium used to complete and close the book. It was a choice dictated by the need to bring life — a posteriori — to a sort of debate on some issues of common interest, shared to support "design alla Sapienza." For these reasons, these subjects turned out to be opportunities of no small importance, for critical assessments in order to reinforce the reasons of our work as teachers and researchers inside and outside design culture. They are useful in measuring a mutual testimony that has been further reinforced over time. What is reported is aggregated around specific themes, treated as key words that invite the reading of the complete essays: the authors' bibliographic references are cited: Vincenzo Cristallo, Federica Dal Falco, Loredana Di Lucchio, Lorenzo Imbesi, Sabrina Lucibello, Raimonda and Riccini.

Naming design

Articulations >   [...] We speak of design when we are referring to specific productive sectors, product design, fashion design, food design, transportation design, lighting design, but also architectural design, urban design [...] We speak of design when we are referring to specific activities within productive processes, interactive design, communication design, eco-design, retail design, brand design [...] And then we come to speak of design whenever there's any form of designing, even in the most disparate fields: sound design, floral design, wedding design [...] In so varied a scenario, one more complicated than complex, what exactly is design, then? And what is its role?

And there it is: whenever we wish to open a reflection on design, we inevitably run into a dual question. The first one, which is not new, is on the most consistent definition of the rules within its action; the second, a more contemporary one (and for this reason perhaps more substantial), of comprehending its external relationships with the cultural, social, and productive system. The first of the two questions — which we may summarize as "What is design?" — has been intrinsic to this discipline since its birth: one need only consider the dichotomy opened by the *Arts & Crafts* movement, or the dissemination work by *Deutscher Werkbund*, or the foundational commitment to the formation of *Bauhaus*, until arriving at the debate triggered in the second half of the twentieth century that attempts to outline the cultural role that Design, through the designing of articles of use, exercises in building social identity. We do not always recall that precisely in Italy, this debate had its starting point and also its most sound points of synthesis.

An initial definition (dating specifically to 1958) of design was offered by Gillo Dorfles, when he wrote that industrial design (and not design!) is (or, rather, was) "that particular category of design   [...] of the objects to be mass-produced using industrialized methods and systems, where the technical side [is combined, ed.] at the outset with an aesthetic element." But the "industrialized" nature of the method as described by Dorfles appealed exclusively to the physical aspects of the object, excluding a priori an additional nature of industrial design which, in another historic definition, this time by Enzo Frateili, was identified as the "theme resulting from the confluence of different disciplines, [capable, ed.] of summarizing in itself the complete production-distribution-consumption cycle."

Therefore, to be able to consider an object the result of the design activity, according to Enzo Frateili it had to include two specific characteristics: seriality, due to the industrial and engineering-based nature of design, and an aesthetic quotient, due to its creative and socio-cultural nature. The definition most consistent with stressing this confluence remains that formulated some years earlier by Tomàs Maldonado, according to which industrial design (and again, not just design) consisted of "coordinating, integrating, and articulating all those factors that, in one way or another, take part in the process constituting the form of the product. And more precisely [Maldonado wrote, ed.] , allusion was made as much to the many factors related to the use, exploitation, and the individual or social consumption of the product (functional, symbolic, or cultural factors) as to those related to its production (technical/economic, technical/construction, technical/systemic, technical/productive,

technical/distributive factors); b) it recombines them in a new form, and c) transfers them again to the outside, in a virtually continuous cycle (if we non do not limit ourselves to considering the action of a single designer); it appears clear, then, that the new information technologies, in their deconstruction of the consequential processes of acquiring and transferring knowledge, modify the society-design-business-society cycle, making it substantially horizontal. We are referring in particular to the open-source phenomenon which, having begun in the IT field, is articulated today in many other cultural and productive settings [...] .
Loredana Di Lucchio, Design on-demand. Evoluzioni possibili tra design, produzione e consumo, in:"Lectures 2," Rdesignpress, 2014

Artefacts >   [...]   In *Artefatti*, Ezio Manzini writes:"the object has always been marked by a dual nature: that of object/prosthesis, which is to say an instrument that amplifies our biological possibilities, and that of the object/sign, a support signifying possible meanings, an integral part of a broader and more complex language of things [...] Today, this binary scheme is not enough; [we are witnessing, ed.] the appearance of a new family of objects capable of quickly performing complex functions [...] as multiplier of cerebral and sensorial activities, [thus defining, ed.] another possible nature of the object [...] ; it was to be the 1990s that consecrated the patinated design of cult objects, of the star system of great designers and of the culture of brand identity." Here, then, is the twenty-first century, opening with an exponential increase in the variety of available products in which, however, the differences appear, on the whole, increasingly irrelevant. It is a multiplication of new elements that do not represent moments of substantial change with respect to their typological precedent, but only a constant mutation that inevitably becomes habit. This phenomenon results in a trivialization of differences, in an instrumental use of novelty without any actual innovative value, in a uniformization of functions and languages. The artificial environment we live in becomes fluid, increasingly fed by the constant transformation of objects that appear newer and newer, while the environment, on the whole, undergoes no change that may be seen as evolution.
It was in those years that Donald Norman, one of the most influential cognitive scientists in recent times, when reflecting on the role of design, stated, in his *The Design of Everyday Things*, that design does nothing more than deal with forms at times so superfluous as to become useless.
But while design continues in this sort of drift, in other fields of knowledge (and of doing), we realize that changes are bringing about a new configuration, in which it is no longer the physical resources that determine the wealth of a social, cultural, and economic system, but the intangible ones: the skills, competence, and above all knowledge of individual persons, of groups, and of the system itself.
The era of the "knowledge economy" is theorized, and the new capital in which to invest is intellectual and reputational capital-that is those kinds of capital that transform knowledge itself from a "given factor" to a "factor to be built." In the form of artefacts, services, and media,

knowledge is contextualized, passing from the places of its production to those of its consumption, and those professions capable of generating and regenerating new knowledge become determinant. This condition defines a landscape, in which there is a proliferation of organizational models (be they businesses, institutions, or social groups) that are no longer univocal, and can be neither repeated nor standardized any longer; they are, rather, changeable models that respond quickly to stimuli (social, cultural, and political), and are thus characterized by high flexibility and adaptability. The clearest consequence of the constant changeability of these organizational models is that the figures that act within them must be just as varied, changeable, and adaptable [...] .
Loredana Di Lucchio, Design on-demand. Evoluzioni possibili tra design, produzione e consumo, in:"Lectures 2," Rdesignpress, 2014

The biography of things > [...] Every object, which can quickly disappear or survive for a long time, has its own life cycle; it has what has been called the *biography of things*.
In the area of archaeology and from the standpoint of philosophy, biology is a recurring metaphor applied to the life of objects. From this perspective, we could study artefacts by following the path laid out between the two extremes: from our initial attraction or resistance to decaying interest. Often, once form is alienated due to the speed of consumption, things disappear; they become useless anti-goods, aged, unusual, and are then reborn after being rediscovered, and their history is questioned. A large quantity of things is left in wills, just as much is broken, thrown into dumps, left in basements, at junk shops, or with antique dealers. The causes, not always decipherable, that lead objects to fall apart are usually connected to the decline of their function of use; " [...] for things that survive and are adopted, there is a sort of metempsychosis corresponding to one or more life cycles." Essentially, things travel, adapt to environments and, if they survive, transmit to us signals of the past that can inform us of their multiple lives [...] The testimonies of the experience of some artefacts described in the text were introduced over time, superimposed upon the original structures: they are historically positioned and contribute towards defining the style of one or more periods, through the traits that mark them [...] Alongside the great stories of important objects, there are the stories of the modest, anonymous objects usually not taken into great consideration [...] The little things that inhabit the world of household practices — irons, brooms, scissors, soaps [...] — were recently studied by Francesca Rigotti and placed in history as real subjects for the purpose of valuing them with respect to the repetition of daily actions. Thus, the minimum systems, if freed of the wear and tear of use, become sources of storytelling: the journeys taken by the pages of a newspaper — from when they are printed until they are used to light an oven or fireplace — or the path that food takes, and all the meanings hidden in the many daily artefacts that go beyond identifying the thing with its mere support. Since time began, it has been commonly held that artefacts are bearers of symbolic nuance, the result of metaphorical projects not ascribable to technical aspects. In this sense, the objects were considered as "active presences" because they were "segments of a human universe" consisting of a complex of hard-to-qualify relationships. According

to this conception, design is to be considered as an activity carried out over time, not only at the service of prosaic practices, but also to convey ritual and emotional elements. [...] With the online revolution and the affirmation of new design practices, this dynamic was accentuated, being implemented in new ways. Disciplinary cross-pollinations seem to have become the only way to achieve innovation, and the types corresponding to emerging life styles are often defined using techniques similar to those of cross breeding, by blending actual and virtual realities. But an umbrella always comes in handy. So there are timeless objects that follow a linear path of evolution. This is particularly evident when it is a matter of artefacts whose value does not lie so much in the decorative and figurative fact, as in the intelligence of the mechanism, which is to say in the most intimate and perfect correspondence between form and function [...] The transmission of knowledge would then take place intermittently, according to the wavelengths that inform us as to permanence of the artefacts, their oblivion, and their evolution, that record with images an era's ideas, tastes, and styles, making reference to personal recollections, but also to myths and events that belong to a culture and a time [...] .

Federica Dal Falco, Design e beni culturali. Giacimenti della modernità, in:"Lectures 3," Rdesignpress, 2014

Creativity/verbal fetish >   [...] In a time of explosive innovation, thinking outside the box is a necessity. And humanistic culture (the humanistic spirit), even in the paradox of its educational decline so clear at the university, is needed by scientists. Da Empoli believes that among the new humanists, those that make the two hemispheres of the brain — art and technology — work at the same time, there are, out of their generosity, designers. But in this case, an anthology of clichés is lost-pleasant for the category of course, but devoted beyond measure to the ephemeral saga of creativity.
In 1957, Gio Ponti made these statements:"Original ideas don't count; in fact, actually original ideas don't exist; [...] ; ideas are received and are re-expressed; ideas are what is reflected in us of a universe of ideas-from past to present, to what we see (some more and some less) in the future with foreboding; [...] people say 'I getting an idea, not 'I'm creating' an idea: etymologically speaking, 'inventing' means finding, not creating." The compendium of these *lectures* cannot lack the word "creativity." Like a nebulous ghost, creativity undermines the dreams and certitudes of those who deal with design in various ways. It is like a spectre that casts no reflection: you will never know its true identity. It is thus imprecise to the point of being tedious. Sometimes irksome, persecutory, and at certain moments sarcastically obscene due to the superficiality with which it is pronounced
 [...] Consequently, there inevitably arises a strong irritation over that sort of twinning imposed upon design by media looking for easily conquered narrative solutions and by a romantic and hagiographic culture proposed by some operators in the sector. If it were possible, I would call a moratorium: prohibiting its use to all for at least five years, while seeing, from a hiding place, the effect this would bring. It is also true that nowadays it is starting to be linguistically replaced, above

all in the world of communication, by the seductive word "smart"; however, its evocative heritage is still present in literary surrogates of every kind, and its mythology dominates the imagery of many [...] Pending suspension of the word, a passage on the Rome underground some months ago restored to me an unexpected measure: a remedy for the improvised debates lying in ambush in the classrooms. The advertising for a school competing against our university degree courses read, word for word, on a 120×180 rotor:"Creativity is nothing without a design" [...] But aphorisms, even if they are cutting and cultured like Oscar Wilde's, withstand time and the flavour of an act of shrewdness. It takes much else to remove the encumbrance of a version, employed in training (in our area), of how articles of use are designed, that places the new generation in contact with their patrimony of creativity. In this way, it "joined the list of expressions emptied by constant misuse, alongside those conceived already empty (see innovation), that are widespread precisely because they are now devoid of content and therefore perfectly generic."

A bulwark must then be built against the demagogy of a mutant creativity in adjectives and nouns used as postulates in design choices. It is a matter of putting up resistance: against a representation of the designer's profession conceived as belonging to those who, under certain conditions, possess the lightning strike of genius that provides magical access to the vision of formal solutions; against certain concept-laden deviations from training, exercised beyond the necessary in the creative drift of such non-trivial words as "service" and "strategy" which, in the darkness of the solutions, appear as therapeutic mantras for exhausted designers [...] . Everyone knows, then, what creativity is, but no one can say what it means precisely except by leafing through a dictionary (the preferred definition: inventive capacity, producing new ideas).

Stefano Bartezzaghi is convinced of this. The well-known puzzle-maker considers it a verbal fetish, vague and oscillating, that in certain circumstances gives itself a paradoxical technical aspect-one more reason enthralling those dealing with the designing of things [...] . It is not admissible, then, that those who manipulate and dispense models of creativity in the form of research applied to designing the artificial-that is, we in the design of academia-do not then run the risk of explaining to our own pupils what it is, what it is not, how to recognize it, how to avoid it, taking it on in explicit form in the paths of training and not submitting to it in hidden configurations inappropriate to our purposes [...] . But if we were to have to undertake the insidious road of teaching, I already nominate one text as indispensable: *Science and Method*, written in 1908 by French mathematician, physicist, astronomer, and philosopher of science Henri Poincaré, who proposes a thinking on creativity that is simple to adopt. He speaks of creativity as "the ability to combine existing elements in new combinations that are useful," and stresses that the empirical, intuitive criterion for deciding the usefulness of the new combination that has just been identified is its beauty. This is a beauty whose assessment, in its turn, presupposes both a specific competence and a sort of emotional sensitivity: it is something that has to do with elegance, harmony, economy of signs, and responding functionally to the purpose. Poincaré's definition is a fertile one, because it is clear and specific, yet universal: it is valid for the sciences, for the arts, for technology [...]

Poincaré's considerations directly regard training: creativity, which is founded upon the preliminary knowledge of the rules to be broken, cannot develop in the absence of skills. Recognizing the importance of being skilled is one of the characteristics of creative personalities, along with curiosity, a need for order and success (not understood in economic terms), independence, dissatisfaction, a critical spirit [...] .
Vincenzo Cristallo, Un si, un no e un tuttavia per il design. La crisi del progetto è per fortuna variabile, in:"Lectures 1," Rdesignpress, 2013

The ethics of design > [...] "Of all the professions, one of the most harmful is industrial design. Perhaps no profession is more false." Papanek, in the guise of designer, philosopher of design, cultural anthropologist, and teacher, comes down hard. With precise, theatrical words, he shoots down industrial designers (whom he also defines as a dangerous breed) and prophesizes a social, ecological, and therefore ethical vision of the world of industrial production that most translate into a sense of responsibility that must accompany every design thought and action for the consequences they have on people and their social setting [...] .Papanek is convinced, in the golden age of mass production in which everything demands to be planned and programmed, that "design has become the most powerful tool with which man shapes his tools and environments [...] This demands high social and moral responsibility from the designer. [...] Design must (therefore, ed.) become an innovative, highly creative, cross-disciplinary tool responsive to the true needs of men. It must be more research-oriented, and we must stop defiling the earth itself with poorly-designed objects and structures." [...] To maintain that industrial design is an instrument capable of concretely satisfying man's needs is only an apparent contradiction. The ideological foundation of all Papanek's work is that of stigmatizing, circumscribing, defining, and limiting, but in the end emancipating the industrial designer's role and work, starting from the consideration that designing, regardless of instruments and content, is at the basis of man's every activity, as the conscious effort to impose meaningful order on his built environment. And, Papanek maintains, it is a significant term that replaces the sound of expressions too rich in semantic implications, like beautiful, ugly, graceful, disgusting, realistic, dark, abstract, and nice. In extreme summary, the ethics Papanek evokes lies entirely in the dimension of the "meaning of design": design is ethical because it is a tool to help the world, and it is ethical for this reason to use it and to allow others to do so, and in the end to teach it [...] .
Enzo Mari, one of the masters of Italian design, to not be misunderstood, when speaking of ethics uses precisely the word "ethics." But first he says:"Utopia is the happy island that is not there, that is not possible. I am a son of the French Revolution, as we all are — Catholics or Jews, Marxists or capitalists. I think the word *égalité* is my faith. I don't believe it is actually possible. But I wish to believe in the possibility of equality through design." And can the word "utopia" be juxtaposed with the word "ethics"? Mari does so, and is believable. He says "it would be worth it to generalize the idea that ethics is the objective of each design (which may be likened to the Hippocratic Oath

[...] What I mean is that all designers should remember the oath every physician makes when graduating from medical school. It may be simplified as follows: the first objective is always to save the patient [...] without ever forgetting to consider the utopia (a new possible model) as the ethical handrail to lean on in order to introduce ethical reforms." Is it rhetoric that makes one's head spin? It is welcome, in spite of so much cheap mediocrity that deceives design. And although it seems more like a fable by Gianni Rodari than from the pen of Mari, here is what he writes to explain himself, saying what I think are the final words on the subject:" [...] when they ask me who is the best designer I know, I always answer: an old farmer who plants a chestnut grove. He knows quite well he will not live long enough to eat its fruits, to warm himself with its wood, or use it to make a stool, or find cool shade beneath its branches in the summertime. He plants it not for himself, but for his grandchildren." For greater balance, as I have done on other occasions, I look for support in Achille Castiglioni, a mentor who with his work introduced me to the civil nature of design. The ethics glimpsed in this nature — never etymologically inconvenienced by the Milanese *industrial designer* — is measured with the principle of doing well, which is equivalent to the very reason for things. Castiglioni, Sergio Polano reminds us, "appears fully aware of the fact that the flow of things, as George Kubler suggests in his *The Shape of Time*, never sees total stoppage: everything that exists today is a replica or variant of something that existed some time ago, and so on, without interruption, back to the dawn of time." It is no accident, Munari said, that, when someone says "I know how to make this thing, too" — the intention here being to not appreciate things simply because they are apparently obvious — they should at least say that they could remake it; otherwise they would have already made it. And this is why, as it is for the historian who cannot clearly separate events of differing natures, even for "an alert designer of artefacts," it will be difficult to act without considering the tradition of replicas that objects possess, before being able to design the next one "towards a non-guaranteed innovation." [...] For Polano, this awareness naturally emerges in the clear love for Castiglioni, for the intelligence that has coalesced over in common things, "for the logical rationality without the redundancy or waste with which utensils are made." But there's nothing to be confused about. The ethical value that Castiglioni shows with his works does not lie in the search for an impersonal and anonymous vetero-socialist sentiment, but rather in having the subjective act of design coincide with the objective processes that historically inform and define things. For Castiglioni, a rationalist free of dogma, it is necessary to investigate the nature of the intimate performance features of artefacts, in order to leave open an ongoing query on their meaning. It is still value that prevails, as well as meaning, the quality of the project that may be seen in the search — Polano maintains — for a "ruthless subtraction and a minimum addition" [...] .
Vincenzo Cristallo, Un sì, un no e un tuttavia per il design. La crisi del progetto è per fortuna variabile, in:"Lectures 1," Rdesignpress, 2013

The crisis of design between the Modern and the Postmodern > [...] The leitmotiv of the word "design" marks the integral theory of the design of the artificial; its breadth is imposed as a metaphor for a unique and enormous global producer of all material and immaterial artefacts; but its meaning, as a programme word, shows a harmful indeterminacy. The entire work of expanding the domain of its meaning that has been done during these years has made this term the logotype of a universal brand most of whose reason and origin is unknown. We are, in my opinion, dealing with a misunderstanding fatal for the discipline, which is to say giving this vocabulary item the burden to comprehend, without any exception whatsoever, the theory of the modernity of contemporary design. All modernity [...] .

The deterioration of the word, and its partial short circuit, are caused, in my opinion, by an even more solid and crucial reason: its postmodernist inclination — a condition not innate but self-determined over time. If industrial design, starting from its terminological genesis, but also from the limits imposed by its historically accredited mandate, has coincided with the principle of modernity, its subsequent articulation in design, in showing itself to be free and inclusive, has been presented as a comprehensive metaphor for the Postmodern. Now that the possibility of comprehending all the transformations of contemporary design in the field of products of use have been exhausted-because the conditions that sustained their ideological and cultural premises have changed-, industrial design has been replaced by design as a set of critical and rhetorical activities capable of containing the possible, thus enabling anyone to be its maker, in accordance with a typically postmodern procedure. Design as a postmodern term can change its appearance and viewpoints, and can as a consequence, demagogically, embody ethical and salvific positions, formalist drifts and drifts of product categories, all at the same time. For many, it is also the term that, through visible and replicable expedients, renders the protean concept of cross-breeding comprehensible-that cross-breeding capable today of overcoming all post-industrial forecasts, because its modern existence is built every day. Only in this way was it possible to overcome the memorable division between what was and is industrially produced, and between what was not and is not in our disciplinary perimeter, the historic contradiction that has traversed the discipline [...] , in modelling itself with respect to demands originating from the ideological and practical culture of industrial design. However, in its postmodernist mandate, design has earned being a factor of the coexistence of designing, and thus offering itself as a context for experimentation between its various configurations in forms, in models, and in new cultural categories. The ambiguity of the postmodern, returning to its kind, lies in the fact that it has kept within itself the movement it intended to overturn, while its decline originates from the circumstance that from being a simple artistic trend, it has evolved to take on political and social meanings in which a preconstituted vision of the world never predominated: every one of its interpretations has therefore equal dignity, also in consideration of the deconstruction of ideologies. This vision, although suggesting a model of democratization of society that has impacted the spread of the pluralism of tolerance, has all the same supported overcoming the myth of objectivity and the primacy of interpretations of facts over the facts themselves.

The real world first became a fable, and then a reality show, the latter nourished by a media populism capable of making one believe anything. The most extreme theory of the Postmodern is

precisely populism, which is to say switching facts and interpretations.
In our field, this disorder — used as a sort of miracle-working expedient — is shown in the affirmation that everything today is design. But this, however much we know it to be true under certain conditions, is not the same as saying that everything that is designed is design. The Postmodern, its populist virus, the disorder of pop imposed between facts and interpretations, has done harm everywhere and fertilized confused ideas, like the one of the necessary counterposition between design and architect re-proposed in many variants and appearances in these past years [...] .
Vincenzo Cristallo, Un si, un no e un tuttavia per il design. La crisi del progetto è per fortuna variabile, in:"Lectures 2," Rdesignpress, 2013

Materials in the culture of design >   [...] Many historical and critical contributions [...] have focused attention on the specific nature of the Italian culture of design, documenting how it has succeeded in affirming itself beyond the traditional technical/engineering culture, so as to in fact build a specific road to the innovation of design, a synthesis in equilibrium between language, form, function, quality of production, and innovation of material.
This unique way of doing Italian design has been constantly renewed, showing itself in particular in certain historical periods.
In the 1950s, Italian design was in its infancy, and it immediately built its own original path of linguistic experimentation linked to materials. The search for an industrial application of Moplen gave rise during these years to Kartell, the company that was to revolutionize daily imagery through a new identity: colourful, light, and democratic.
In the 1960s, Andrea Branzi's "primary design" shifted attention to soft qualities and the expressive/sensorial identity of materials, broadening the competence of design until reinventing their surface. Thus was born the "design of materials" as we understand the term, to the point that even the business world, led by Montefibre, addressed such apparently minor productive problems as colour, decoration (particularly of fabrics), surface, light, and sound, placing them at the centre of sophisticated industrial strategies.
In the late 1980s, design was projected towards the "technological unexpected," which is to say towards a world in which everything is possible, and in which materials, with new and extraordinary performance features, were pressed more than they ever had before to the limits of their technical and symbolic possibilities, but also drawn upon and transferred by other productive sectors.
And so it was until the 1990s, years during which speaking of materials meant knowing and governing a complexity and a hyper-choice [...] in fact no longer manageable by the designer or even less so by the technician. These were also the years during which the complete miniaturization and dematerialization of the world was prophesied: a catastrophe that was foretold but had never taken place. And thus we come down to our own days, when the perfection of methods, and design strategies, are constantly changing the landscape of reference, first of all by amplifying the potentials

of the design of materials, for which everything seems possible, but also and above all by redefining, with Antonio Petrillo, a "new order of artificiality."

And therefore, the new technologies are no longer used only to personalize, to extend, and to modify the physical properties of the material, but as an instrument to transform their sensory characteristics, guiding the response of our senses a priori. At the same time, nanotechnologies, by making it possible to design at the macro, micro, and even nano level, tend to make the objects more "simple" on the surface, and more complex in their DNA.

This is a genuine design revolution, which we will call "hyper-design" and that is making it possible to work in parallel on the subject, as well as on form/function, opening new and infinite paths in the search for the supermaterial: light, sensory, high-performance, sustainable, magic, and — why not? — alive [...] Therefore, if on the one hand it is technology and science that inspire us, by providing us with supermaterials endowed with genuine superpowers, on the other it is the approach we define as "creative" that, also in the absence of resources in terms of raw materials or money, can generate innovation at both the product level and at the level of material. This is particularly the case of Italian design which, although unable to rely on major investment and raw materials, has still made creative use of materials, giving voice to innovation through new products.

Thus, "creative thought" has arrived at imagining what is not there, but that might only be, generating experimentation efforts close both to the dynamics inherent to the art world and to those of sophisticated space technologies, thereby giving life to kaleidoscopes of possibilities that have changed artefacts, materials and uses in order to give rise to revolutions great and small [...] . With nanotechnologies, the material and its structures may be built atom by atom, molecule by molecule [...] ; materials can be programmed to reiterate natural processes like the lotus effect, which is to say the autonomous self-cleaning ability observed chiefly in lotus flowers; or like the "super-adherence" which, by imitating the gecko's ability to maintain perfect adhesion to any kind of surface (even in a void and underwater), reproduces the unique conformation of the tissue beneath its feet.

This gives rise to interesting experimentations on fabrics, like the one developed by Ermenegildo Zegna with Traveller, a stain-resistant fabric made with a new type of finish inspired by the Lotus effect, and that does not alter the fabric's performance characteristics and softness [...] . And then there are the "genetically modified" natural materials, as is the case [...] of "bio-LED" which in the near future might be used, for example, to light roads at night, transforming trees into colourful streetlamps [...] or having them take on all the colours of the rainbow depending on their nano-dimensions [...] Nano-dyes can dye without having to fix the colours derived from synthetic pigments, and therefore without using chemical and surfactant mordants [...] ; the iridescence of colours can be designed, reproducing the natural one of the wings of butterflies or other insects.

The colour of butterfly wings, in fact, depends not so much on pigments as, as is now well known, on special micro-structures present in the wings, that reflect and refract the light incident upon their surface [...] . The reproduction of natural photonic structures opens interesting scenarios for perfecting industrial systems able to prevent the counterfeiting of banknotes, credit cards, and other documents [...] . We are in the field where progress is being made in biomimetics, a science for which nature is not just a wealth of harmonious forms to be imitated, but an infinite register of sensitive and sustainable structures that use less matter and energy, in addition to sophisticated natural processes, in order to be highly efficient [...] . This also gives rise to new scenarios for sustainability [...] — scenarios according to which scraps and "material waste" would be transformed into nutrients of use for the growth of organism-objects, just as they are for biological ones [...] Thanks to the work of designers like Lehanneur, Revital Cohen, and Susana Soares, who in 2008 took part in the exhibition curated by Paola Antonelli, "Design and the Elastic Mind," at MoMA, SymbioticA has developed certain products that incorporate living beings, generating interactive and sensitive profiles of colour [...] Of interest [...] is the Jelte van Abbema's Symbiosis design which, by cultivating bacteria and cells, has made a poster using typographical characters in continuous and vital evolution, because they are based on the life cycle of multiplication and death of *Escherichia Coli* bacteria on cellulose and agar (a polysaccharide obtained from red algae).

Also interesting are some experiments to make an artificial material similar to cellulose, Xylinum, named for the bacterium that produces it: feeding on sugar, the bacterium fabricates a 100% biodegradable cellulose fibre structure [...] whose properties may also be regulated by changing its genetic code.

In a kind of reverse process, technological innovations return us to nature which can be imitated in form-increasingly organic thanks to articulated productive and information technology processes capable of governing complex meshes-, in behaviour, more and more naturally alive and intimately able to sustain itself, and in infinite variety [...] .

Sabrina Lucibello, L'approccio creativo ai materiali per il design: tra supermateriali e iperprogettazione, in:"Lectures 2," Rdesignpress, 2014.

The word "type" for design

They are called "types" >   [...] Andrea Branzi, who is an advocate of the evolutionary history of design, uses a highly effective image: from prehistory to the contemporary era, there has been a single flow of artefacts in transformation and, link after link, objects move together (perhaps dance together) in the same amniotic fluid constituted by anthropological components and the traditions of the past. The coexistence of populations of objects due to uninterrupted transformations, between archaic categories and emerging forms, urges us to work imaginatively. For example, by imagining we are adopting the evolutionary scheme to study the life of things, we could group together homogeneous classes of artefacts in galleries of extinct, surviving, and evolved specimens; from the most primitive placed at one end of the initial series, to the most evolved at the other end.
In design, industrially produced objects are called types when they are the same as each other and identical to a model that must be conceived as already completed at the very act of its production, and not be submitted to subsequent manipulations modifying its appearance. This classical definition clearly divides the world of objects between artisinal and industrial, between before and after the advent of the machine as a multiplier of infinite series, even if it includes, within the industrial design hybrids, objects made with mixed techniques such as industrially produced wooden furniture, whose finishing is done by hand. But, at the moment when we pose the problem of cataloguing artefacts, a fundamental theme crops back up: design's relationship with history.
The premise upon which the classification system recently proposed by De Fusco is based starts from the consideration that the autonomous origin of industrial design does not derive from parthenogenesis, but is a highly complex phenomenon connected with an ancient genealogy: products unchanged over time, transformations of forms but not of substance, the simultaneous presence of an artisanal dimension and of technical reproduction.
Industrial design is a unitary concept in terms of its general phenomenology (from design to consumption), but its breakdown into families cannot help concerning the product category fields with respect to which it is fragmented into a variety of material products [...] .
Federica Dal Falco, Design e beni culturali. Giacimenti della modernità, in:"Lectures 3," Rdesignpress 2014

The distinctive traits of objects >   [...] Perhaps we truly are, as the architectural historian Peter Carl says, in the fourth phase of the theory of typology since it was elaborated during the Enlightenment. Although Carl has architecture in mind, it is not out of place to turn his observation to design as well. I would like to try to do this by retracing — at least in their general features — the historical phases of the history of type, and its concrete role in designing artefacts [...] In the face of the evident upset of the landscape of the objects of our material world, in the presence of technologies that promise radical transformations in the way of thinking, designing, making, and using artefacts old and new, we must admit that, of the many strained certainties of the twentieth century, we are also about to be abandoned by the one linked to the concept of type, in all its innumerable facets. Never more than today, the material scaffolding underpinning the modern world is changing its characteristics from its depths. Never more than today, we are noting that the widespread use of digital and integrated technologies, of smart materials, of design software- including open source software-is plotting against the central importance of the problem of form in the world of products. This results in the consequent loss of rank of the historical form/function dichotomy [...] .

If we ask ourselves about what the word "type" has meant for design, it is presented as a two-faced Janus. On the one hand, it is the founding exemplar of the new way of designing and producing objects in the modern age (typing, or standardization), and on the other it is an essential element for recognizing these same objects on the social scene (typology). Therefore, one interpretation key is historic and the other operative; one is evolutionary and the other productive; one regards products, and the other regards the act of designing. This would be enough to comprehend its importance for design culture. Despite this, the typological question did not, in the historical or theoretical pages specifically of design, play a similar leading role, or at least — at appears to me — it was not equal to its objective relevance in the dynamics between design/production/consumption. While, in the context of architecture, type and its role in the culture of building were at the centre of several different theoretical propositions and critical analyses, design culture (particularly Italian design culture) in fact inherited its arrangement, but without being able to develop it autonomously. Actually, terms like type, typology, and typing were relegated rather quickly to the toolbox of modernism, as a narrow heritage of industrialization and a deleterious tool of information. The fundamental role played by object typologies in the context of our material culture and in our operative and functional relationship with the real (and virtual) world has too often been forgotten. There is no doubt that the typologies, and the ways of typing we use to develop them, are the main instrument of the production and reproduction of the material world, but also an essential element for giving meaning to our place in it [...] Looking back with the long view of the ethnographer, we may think of the artefacts of our material culture as a matter of proceeding — at times slowly, at times quite quickly — between variants and invariants. Such archetypical forms as the bowl, the hammer, the knife, the stool, the basket, and the vase, to name but a few, represent an original typological catalogue that, over time, has been developed and enriched with the innumerable cultural variants made in the different societies. But things become complicated when, on the threshold of modernity, alongside these faithful companions originating from the primordial technical activity of *homo faber*, new products and technical objects appear in greater and greater numbers, the offspring of the Industrial Revolution and new protagonists in the material culture of modernity [...] .

Raimonda Riccini, Il senso del design per il tipo, in:"Type & Model | idee, progetti, azioni," Planning, Design, Tecnology Journal no.4, Rdesignpress, 2015

Type and Modern Movement >   [...] Among the tactics and strategies of an incipient modernism, that of typing was above all a need with which the mass production system had to come to terms, something the leaders of the Modern Movement were fully aware of. One example for all is Gustav Platz. His book, *L'architettura della nuova epoca*, written in 1927 as an attempt to summarize production and design thinking in the first three decades of the twentieth century, expressed the widely shared point of view: "mass production must resort to types and norms. All products, from cloths to household utensils, from the simplest article of use to the constituent parts of a building, are made with the help of the machine [...] The definition of types, however, is not only a consequence of the mechanization of the productive process, but also derives from the repetition of typical needs, and thus constitutes the conscious demand of every economically correct social policy" [...] . Writing in 1955, Walter Gropius returns with the same emphasis on the theme of the development of standard types: "the creation of standardized types for articles of daily use is a social necessity. The standard product is not at all an invention of our time: only the methods to produce it have changed. Even today, it applies the highest level of civilization, the search for the more perfect type, the separation of what is essential and over-personal from what is personal and accidental" [...] While, during the first phase of capitalism, typing had functioned as a system to bring order to production and to products, and to regulate the proliferation of new types of objects, now, as a response to market turbulence, the rigidities typical of the Fordist model are loosening, above all through a technological acceleration (robotization) and a different organizational logic ("just in time") that focuses on demand and thus the consumer. Without retracing the steps that accompanied this transformation, from "Toyotism" on, we can say, however, that we are dealing today with a "manufacture in transition," by virtue precisely of the leading role played by increasingly pervasive technologies, and the emergence of the figure of the consumer/designer/producer [...] .

Raimonda Riccini, Il senso del design per il tipo, in: "Type & Model | idee, progetti, azioni," Planning, Design, Tecnology Journal no. 4, Rdesignpress, 2015

Design: new paradigms

Design in the Open Source era > [...] We are now seeing a blurring of the boundaries between different paradigms, the advent of the open-source and free software era, in which specialization gives way for the trans-disciplinary and for holistic and integrated approaches, and in which the Mephistophelean solitude of the artist and designer in the creative process becomes an act of co-creation in an open and collaborative process. Design and the typing processes of forms teach us, in practice, how different challenges equate to different solutions, where there is no right or wrong methodology, but an attitude that casts doubt on the phenomena and on the relationship between problems, solutions, and contexts. In naming "design," we associate such categories as fluidity, connection, the passage from the vertical to the diagonal, from top-down to bottom-up, from content protection to open-source sharing, dissemination before identity-based obsession; but also the constellations and openness of meanings, the added value provided by emotions as a complement to a purely rational world, the value of ambiguity and the wealth of sensory experiences, the blurring between the real and the virtual, and the return to a focus on the pre-rational state, where objects may be seen in relation to subjects. We may speak of relational objects, a sort of mature appendix to Winnicott's transitional objects, as an "in-between" space dynamically shaped by tensions of energy, in which a pure object is not designed-but also its relationship with the environment, the user's experience, the touchpoints of the service, a transformation process, and so on. Design takes on the burden of literally designing the interstitial space that can open a gateway for a new paradigm, a new episteme, a new way of thinking [...] In 1958, in seeking archetypal forms from elementary industrial processes, Enzo Mari designed for Danese a limited-edition tray called "putrella" or "beam." Through a process of de-functionalization and re-functionalization in a different setting typical of the ready-made, the designer reused a section of double-T beam, bending it at the ends to hint at the shape of a tray. The product challenges the bourgeois home and its object universe, using a semi-processed piece typical of the construction site. Likewise, it provides an ironic take on functionality and resistance, precisely by bending it at the ends. Like Barthes in the analysis of language, and like many other radical designers of his time, Mari uses forms and types to challenge their rules of function and manipulate their meanings: with the approach of a hacker, he comprehends and emphasizes that convergence of technological factors that enabled the production of a type, and reprocesses them together to construct new meanings and affirm yet again a dynamic vision of the typing processes of forms.
Many years later, history repeated itself by showing how this can be an open and infinite process: In 2014, Ronen Kadushin published his files for the laser manufacturing of a "putrella" tray, which was redesigned to be shared in open source, and thus produced and assembled directly by the final user. From a double-T metal profile, the design became a two-dimensional drawing produced on sheets of paper to be put together as a pre-cut pattern; it has therefore now lost, for good, the original qualities of mechanical resistance to support construction elements. Once again, as Simondon observed, technology plays an important role in mediating the archetypal forms with the physical and social dimension of the world, contributing towards 'opening' the typologies in order to turn

them into a patrimony [...] .
Lorenzo Imbesi, Il Collezionista, il Designer e l'Hacker, in:"Type & Model | idee, progetti, azioni," Planning, Design, Tecnology Journal no. 4, Rdesignpress, 2015

The design of desktop manufacturing > [...] The first phenomenon to which I wish to draw attention is the exponential spread in recent years of those technologies that we may group together under the label of "Rapid Manufacturing."
Although Rapid Manufacturing-and with it, Rapid Prototyping — are not brand — new processes (they date to the 1990s), their accessibility has grown in recent years, to the point that the approach methodologies are still multiple and differ greatly from one another, because they are still in the trialling phase. To summarize, the term "Rapid Manufacturing" refers in particular to those additive processes that allow objects to be built by adding material starting solely and directly from the mathematical model made in 3D CAD: a binary matrix that tells the machine where to deposit the material, where to strike it with the laser beam, where to melt it. Considering that a) the productive process that is determined is no longer necessarily serial; that b) these technologies allow geometries to be handled with no limits in form (for now, the only limits are related to size) and that c) materials can be used that are specially designed to best perform the function of the finished product without producing scrap while — potentially — being easy to recycle, the picture becomes certainly innovative. But beyond the technological and economic opportunities, what is interesting to analyze here is the substantial change these technologies enable in the process of defining the product (and also its consumption).
We have already seen that one of the characteristics that has distinguished, and still distinguishes, the figure of the designer and that in actuality justified his professional and disciplinary birth, is that of being, within the mass industrial process, the one who, upstream, defines the result of the process (the product) and transfers it through specific information to all the figures that variously operate in the same process, but without intervening for this reason directly in the realization phases. In other words, the designer represents that part of the craftsman that had been removed from the industrial worker. Instead, thanks to Rapid Manufacturing, the designer reappropriates the knowledge of the process, managing it directly through the 3D modelling of the article, and reverts to a condition more like that of the pre-industrial craftsman. Thus, albeit mediated by the digital tool, the distance between conceiving and producing is minimal, the relationship is direct and immediate, and what is designed is already the article's final form [...] .
Loredana Di Lucchio, Design on-demand. Evoluzioni possibili tra design, produzione e consumo, in: "Lectures 2," Rdesignpress, 2014

Design on demand > [...] The term "open-source" came into being in the late 1990s to replace the term "free-software" and give greater emphasis not to the product (the free software, to be precise) but to the process that was triggered starting from an open IT "source," which everyone could access and everyone could implement, edit, or upgrade. What was activated was and is a different "creative" process in which a potentially infinite network of players share, process, and disseminate knowledge, whether intangible (we are thinking of the experiences of CopyLeft, or of the Creative Commons for the cancellation of copyright)-or tangible, and thus regarding products that are no longer the result of the design capacity of a single party or a single company. It is this last articulation of open source that has considerably impacted design, to the point that we speak today of "Open Design," where what disappears is the "ownership" by the individual party of a new product's development process, given the acquisition of the infinite potential that comes from knowledge of it [...] The outlines of a significant evolution of the relationship between design, production, and consumption, or better, an evolution of the very subjects of the relationship, which are partially (and at times wholly) changing their appearance, their skills, but also their very objectives, are expressed around three key concepts: a) desktop manufacturing; b) total control, c) the smart supply chain. What is most interesting in my opinion is that simultaneous presence of these concepts shifts the focus both of production (which is more understandable), but also and above all of design and consumption (a particularly new element) from product innovation to a different idea of process innovation. We know that product innovation focuses attention on the results of the productive processes and, only as consequence, on the processes themselves; in the case of Italian design, this has taken on particular importance by attributing the same index of innovation to the product's semantic cultural values. We also know that process innovation, on the other hand, by referring precisely to "operative processes," focuses on the roles and instruments of all the subjects participating in the entire design/production/consumption cycle. In this case, however, process innovation is dynamic because, as we have seen, the development and spread of Rapid Manufacturing technologies have opened the way for a logic of process that entails a transition from the unique, specialist knowledge of the few to synergy among different kinds of knowledge. For Design, but also for business, this means overcoming the phenomenon of the design firm and making the most of the network [...] The spread of information technologies has shifted the focus of the creative process from mere conception (the design act, in fact) to the conception/production dualism, which is quite close to "Do-It-Yourself" (or, better, to its updated version:"Do It With Others"), in which the designer in fact becomes a maker and has the opportunity to take total control over the process itself, and in this way implements continuous experimentation rather than waiting for the "customer"; the designer goes from being an output processor to an input supplier; lastly, the introduction of digital into the production management processes, even in traditional sectors (like Italian-made manufacturing) is transforming the supply chain from a merely problem-solving path to a "smart" pattern, reactive and flexible to a variety of external inputs. A paradigmatic change is thus implemented, that goes from the strategy (of a modernist matrix) of developing new knowledge only after having reduced the phenomena to homogeneous units (specializations), to the more properly postmodern one that instead arrives at knowledge by relating the phenomenon to the qualitatively broad context. It is a context where the key lies in the network

or, to use D. Weinberger's metaphor, in the "room": it is therefore no longer important how much one of the subjects "in the room" is prepared, because it will always be the room that is the smartest, since it expresses the sum — or better, the overlapping — of everyone's ideas, debates them, explains them, and applies them to new contexts.
The design/production/consumption relationship thus transitions, using a communications metaphor, from a condition of "broadcasting" — which is one-way (from one point to all points) — a specific content (individual design skills offered to provide a response to the needs of a single business and social reality), to being an on-demand condition, which is above all interactive (from one point to one point), where content and skills are flexible and conform to the interlocutor's needs [...] .

Loredana Di Lucchio, Design on-demand. Evoluzioni possibili tra design, produzione e consumo, in:"Lectures 2" Rdesignpress, 2014

图书在版编目（CIP）数据

设计: 文本与语境: 汉英对照 /（意）托尼诺·帕里斯著；林晶晶译. —上海: 华东师范大学出版社, 2018

ISBN 978-7-5675-8510-2

Ⅰ.①设⋯ Ⅱ.①托⋯ ②林⋯ Ⅲ.①设计学—汉、英 Ⅳ.①TB21

中国版本图书馆CIP数据核字（2018）第255583号

设计——文本与语境

著　　者	托尼诺·帕里斯
译　　者	林晶晶
策划编辑	阮光页
责任编辑	朱妙津　谢　莹
审读编辑	李贵莲
责任校对	邱红穗　王丽平
装帧设计	陈文皓

出版发行	华东师范大学出版社
社　　址	上海市中山北路3663号 邮编 200062
网　　址	www.ecnupress.com.cn
电　　话	021-60821666　行政传真 021-62572105
客服电话	021-62865537　门市（邮购）电话 021-62869887
地　　址	上海市中山北路3663号华东师范大学校内先锋路口
网　　店	http://hdsdcbs.tmall.com

印刷者	上海盛通时代印刷有限公司
开　本	787×1092　16开
印　张	16
字　数	320千字
版　次	2018年11月第1版
印　次	2018年11月第1次
书　号	ISBN 978-7-5675-8510-2/J·375
定　价	58.00元

出版人　王　焰

（如发现本版图书有印订质量问题，请寄回本社客服中心调换或电话021-62865537联系）